Math Made Nice-n-Easy Books™

In This Book:

- ## Integration Formulas

- ## Combinations & Permutations

- ## Probability

"MATH MADE NICE-n-EASY #9" is one in a series of books designed to make the learning of math interesting and fun. For help with additional math topics, see the complete series of *"MATH MADE NICE-n-EASY"* titles.

Based on U.S. Government Teaching Materials

Research & Education Association
61 Ethel Road West
Piscataway, New Jersey 08854

Dr. M. Fogiel, Director

MATH MADE NICE-N-EASY BOOKS™
BOOK #9

Printed in the United States of America

Library of Congress Control Number 2001091428

International Standard Book Number 0-87891-208-8

MATH MADE NICE-N-EASY is a trademark of Research & Education Association, Piscataway, New Jersey 08854

WHAT "MATH MADE NICE-N-EASY" WILL DO FOR YOU

The "Math Made Nice-n-Easy" series simplifies the learning and use of math and lets you see that math is actually interesting and fun. This series of books is for people who have found math scary, but who nevertheless need some understanding of math without having to deal with the complexities found in most math textbooks.

The "Math Made Nice-n-Easy" series of books is useful for students and everyone who needs to acquire a basic understanding of one or more math topics. For this purpose, the series is divided into a number of books which deal with math in an easy-to-follow sequence beginning with basic arithmetic, and extending through pre-algebra, algebra, and calculus. Each topic is described in a way that makes learning and understanding easy.

Almost everyone needs to know at least some math at work, or in a course of study.

For example, almost all college entrance tests and professional exams require solving math problems. Also, almost all occupations (waiters, sales clerks, office people) and all crafts (carpentry, plumbing, electrical) require some ability in math problem solving.

The "Math Made Nice-n-Easy" series helps the reader grasp quickly the fundamentals that are needed in using math. The reader is led by the hand, step-by-step, through the various concepts and how they are used.

By acquiring the ability to use math, the reader is encouraged to further his/her skills and to forget about any initial math fears.

The "Math Made Nice-n-Easy" series includes material originated by U.S. Government research and educational efforts. The research was aimed at devising tutoring and teaching methods for educating government personnel lacking a technical and/or mathematical background. Thanks for these efforts are due the U.S. Bureau of Naval Personnel Training.

Dr. Max Fogiel
Program Director

Contents

Chapter 17
Probability

Chapter 15
Integration Formulas

In this chapter several of the integration formulas and proofs are discussed and examples are given. The formulas in this chapter are basic and should not be considered a complete collection of integration formulas. Integration is so complex that tables of integrals have been published for use as reference sources.

In the following formulas and proofs, u, v, and w are considered functions of a single variable.

Power of a Variable

The integral of a variable to a power is the variable to a power increased by one and divided by the new power.

Formula:

$$\int x^n \, dx = \frac{x^{n+1}}{n+1} + C, \; n \neq -1$$

Proof:

$$d\left(\frac{x^{n+1}}{n+1} + C\right) = \frac{(n+1)x^{n+1-1}}{(n+1)} \, dx$$

$$= \frac{(n+1)x^n}{(n+1)} \, dx$$

$$= x^n \, dx$$

therefore

$$\int x^n \, dx = \frac{x^{n+1}}{n+1} + C, \ n \neq -1$$

EXAMPLE: Evaluate

$$\int x^5 \, dx$$

SOLUTION:

$$\int x^5 \, dx = \frac{x^{5+1}}{5+1} + C$$

$$= \frac{x^6}{6} + C$$

EXAMPLE: Evaluate

$$\int x^{-5} \, dx$$

SOLUTION:

$$\int x^{-5} \, dx = \frac{x^{-4}}{-4} + C$$

Product of Constant and Variable

When the variable is multiplied by a constant, the constant may be written either before or after the integral sign.

Formula:

$$\int a \; du = a \int du = au + C$$

Proof:

$$d(au + C) = a \; d \left(u + \frac{C}{a}\right)$$

$$= a \; du$$

therefore

$$\int a \; du = a \int du = au + C$$

EXAMPLE: Evaluate

$$\int 17 \; dx$$

SOLUTION:

$$\int 17 \; dx = 17 \int dx$$

$$= 17x + C$$

EXAMPLE: Evaluate

$$\int 3x^4 \; dx$$

3

SOLUTION:

$$\int 3x^4 \, dx = 3 \int x^4 \, dx$$

$$= (3) \frac{x^5}{5} + C$$

$$= \frac{3x^5}{5} + C$$

Sums

The integral of an algebraic sum of differentiable functions is the same as the algebraic sum of the integrals of these functions taken separately.

Formula:

$$\int (du + dv + dw) = \int du + \int dv + \int dw$$

Proof:

$$d(u + v + w + C) = du + dv + dw$$

therefore

$$\int du + \int dv + \int dw = u + C_1 + v + C_2 + w + C_3$$

where

$$C_1 + C_2 + C_3 = C$$

Then

$$\int du + \int dv + \int dw = u + v + w + C$$

and

$$\int (du + dv + dw) = \int du + \int dv + \int dw$$

$$= u + v + w + C$$

EXAMPLE: Evaluate

$$\int (3x^2 + x)\, dx$$

SOLUTION:

$$\int (3x^2 + x)\, dx = \int 3x^2\, dx + \int x\, dx$$

$$= x^3 + C_1 + \frac{x^2}{2} + C_2$$

$$= x^3 + \frac{x^2}{2} + C$$

EXAMPLE: Evaluate

$$\int (x^5 + x^{-3})\, dx$$

SOLUTION:

$$\int (x^5 + x^{-3})\, dx = \int x^5\, dx + \int x^{-3}\, dx$$

$$= \frac{x^6}{6} + C_1 + \frac{x^{-2}}{-2} + C_2$$

$$= \frac{x^6}{6} - \frac{x^{-2}}{2} + C$$

$$= \frac{x^6}{6} - \frac{1}{2x^2} + C$$

Practice Problems

Evaluate the following integrals:

1. $\int x^6 \, dx$

2. $\int x^{-4} \, dx$

3. $\int 17x^2 \, dx$

4. $\int \pi r \, dr$

5. $\int 7x^{1/2} \, dx$

6. $\int (x^7 + x^6 + 3x^3) \, dx$

7. $\int (6 - x^3) \, dx$

Answers

1. $\frac{x^7}{7} + C$

2. $-\frac{1}{x^3} + C$

3. $\frac{17}{3}x^3 + C$

4. $\frac{\pi r^2}{2} + C$

5. $\frac{14}{3} x^{3/2} + C$

6. $\frac{x^8}{8} + \frac{x^7}{7} + \frac{3}{4}x^4 + C$

7. $6x - \frac{x^4}{4} + C$

Power of a Function of x

The integral of a function raised to a power and multiplied by the derivative of that function is found by the following steps:

1. Increase the power of the function by 1.

2. Divide the result of step 1 by this increased power.

3. Add the constant of integration.

Formula:

$$\int u^n \, du = \frac{u^{n+1}}{n+1} + C, \ n \neq -1$$

Proof:

$$d\left(\frac{u^{n+1}}{n+1} + C\right) = \frac{(n+1)u^n}{n+1} \, du$$

$$= u^n \, du$$

therefore

$$\int u^n \, du = \frac{u^{n+1}}{n+1} + C$$

NOTE: Recall that

$$\frac{d}{dx}\left[\frac{2}{3}(2x-3)^3\right] = (3)(\frac{2}{3})(2x-3)^2$$

$$= 2(2x-3)^2$$

EXAMPLE: Evaluate

$$\int(2x-3)^2(2)\,dx$$

SOLUTION: Let

$$u = (2x-3)^-$$

and

$$du = 2\,dx$$

and

$$n = 2$$

Then

$$\int u^n\,du = \frac{u^{n+1}}{n+1} + C$$

$$= \frac{u^3}{3} + C$$

and by substitution

$$\int(2x-3)^2(2)dx = \frac{(2x-3)^3}{3} + C$$

When using this formula the integral must contain precisely du. If du is not present it must be placed in the integral and then compensation must be made.

EXAMPLE: Evaluate

$$\int (3x + 5)^2 \, dx$$

SOLUTION: Let

$$u = (3x + 5)$$

and

$$du = 3 \, dx$$

We find dx in the integral but not 3 dx. A 3 must be included in the integral in order to fulfill the requirements of du.

In words, this means the integral

$$\int (3x + 5)^2 \, dx$$

needs du in order that the formula may be used. Therefore, we write

$$\frac{3}{3} \int (3x + 5)^2 \, dx$$

and recalling that a constant may be carried across the integral sign, we write

$$\frac{3}{3} \int (3x + 5)^2 \, dx = \frac{1}{3} \int (3x + 5)^2 \, 3 \, dx$$

Notice that we needed 3 in the integral for du and we included 3 in the integral, then compensated for the 3 by multiplying the integral by 1/3.
Then

$$\frac{1}{3} \int (3x + 5)^2 \, 3 \, dx = \left(\frac{1}{3}\right) \frac{(3x + 5)^3}{3} + C$$

$$= \frac{1}{9}(3x + 5)^3 + C$$

EXAMPLE: Evaluate

$$\int x(2 + x^2)^2 \, dx$$

SOLUTION: Let

$$u = (2 + x^2)$$

and

$$du = 2x \, dx$$

Then

$$\int x(2 + x^2)^2 \, dx = \frac{2}{2} \int x(2 + x^2)^2 \, dx$$

$$= \frac{1}{2} \int 2x(2 + x^2)^2 \, dx$$

$$= \left(\frac{1}{2}\right) \frac{(2 + x^2)^3}{3} + C$$

$$= \frac{(2 + x^2)^3}{6} + C$$

Practice Problems

Evaluate the following integrals:

1. $\int (x^2 + 6)(2x)\ dx$
2. $\int x^2 (7 + x^3)^2\ dx$
3. $\int (3x^2 + 2x)^2 (6x + 2)\ dx$
4. $\int (6x^3 + 2x)^{1/2} (9x^2 + 1)\ dx$
5. $\int (x^2 + 7)^{-2}\ x\ dx$

Answers

1. $\dfrac{(x^2 + 6)^2}{2} + C$

2. $\dfrac{(7 + x^3)^3}{9} + C$

3. $\dfrac{(3x^2 + 2x)^3}{3} + C$

4. $\dfrac{(6x^3 + 2x)^{3/2}}{3} + C$

5. $\dfrac{1}{2(x^2 + 7)} + C$

11

Quotient

In this section three methods of integrating quotients are discussed but only the second method will be proved.

The first method is to put the quotient into the form of the power of a function. The second method results in operations with logarithms. The third method is a special case in which the quotient must be simplified in order to use the sum rule.

Method 1

If we are given the integral

$$\int \frac{2x}{(9 - 4x^2)^{1/2}} \, dx$$

we observe that this integral may be written as

$$\int 2x(9 - 4x^2)^{-1/2} \, dx$$

and by letting

$$u = (9 - 4x^2)$$

and

$$du = -8x \, dx$$

the only requirement for this to fit the form

$$\int u^n \, du$$

is the factor for du of -4. We accomplish this by multiplying 2x dx by -4, giving -8x dx which is du. We then compensate for the factor -4 by multiplying the integral by - 1/4.

Then

$$\int \frac{2x}{(9 - 4x^2)^{1/2}} \, dx = \int 2x(9 - 4x^2)^{-1/2} \, dx$$

$$= -\frac{1}{4} \int (-4) (2x)(9 - 4x^2)^{-1/2} \, dx$$

$$= -\frac{1}{4} \int -8x(9 - 4x^2)^{-1/2} \, dx$$

$$= (-\frac{1}{4}) \frac{(9 - 4x^2)^{1/2}}{\frac{1}{2}} + C$$

$$= -\frac{(9 - 4x^2)^{1/2}}{2} + C$$

EXAMPLE: Evaluate

$$\int \frac{x}{(3 + x^2)^{1/2}} \, dx$$

SOLUTION: Write

$$\int \frac{x}{(3 + x^2)^{1/2}}\ dx = \int x(3 + x^2)^{-1/2}\ dx$$

Then, let

$$u = (3 + x^2)$$

and

$$du = 2x\ dx$$

The factor 2 is used in the integral to give du and is compensated for by multiplying the integral by 1/2.
Therefore

$$\int x(3 + x^2)^{-1/2}\ dx = \frac{1}{2} \int 2x(3 + x^2)^{-1/2}\ dx$$

$$= (\frac{1}{2}) \frac{(3 + x^2)^{1/2}}{\frac{1}{2}} + C$$

$$= (3 + x^2)^{1/2} + C$$

Practice Problems

Evaluate the following integrals:

1. $\int \dfrac{x}{(2 + x^2)^{1/2}}\ dx$

2. $\int \dfrac{dx}{\sqrt[3]{3x + 1}}$

3. $\int \dfrac{dx}{(3x + 2)^5}$

Answers

1. $(2 + x^2)^{1/2} + C$

2. $\dfrac{(3x + 1)^{2/3}}{2} + C$

3. $\dfrac{-1}{12(3x + 2)^4} + C$

Method 2

In the previous examples, if the exponent of u was -1, that is

$$\int u^n\ du$$

where

$$n = -1$$

we would have applied the following.

15

Formula:

$$\int \frac{du}{u} = \ln u + C, \ u > 0$$

Proof:

$$d(\ln u + C) = \frac{1}{u} \ du$$

therefore

$$\int \frac{du}{u} = \ln u + C$$

EXAMPLE: Evaluate the integral

$$\int \frac{1}{x} \ dx$$

SOLUTION: If we write

$$\int \frac{1}{x} dx = \int x^{-1} \ dx$$

we find we are unable to evaluate

$$\int x^{-1} \ dx$$

by use of the power of a variable rule so we write

$$\int \frac{1}{x} dx = \ln x + C$$

because the 1 dx in the numerator is precisely du and we have fulfilled the requirements for

$$\int \frac{du}{u} = \ln u + C$$

EXAMPLE: Evaluate

$$\int \frac{2}{2x + 1} \, dx$$

SOLUTION: Let

$$u = 2x + 1$$

and

$$du = 2 \, dx$$

then we have the form

$$\int \frac{du}{u} = \ln u + C$$

therefore

$$\int \frac{2}{2x + 1} \, dx = \ln (2x + 1) + C$$

EXAMPLE: Evaluate

$$\int \frac{2}{3x + 1} \, dx$$

SOLUTION: Let

$$u = 3x + 1$$

and

$$du = 3 \, dx$$

17

We find we need 3 dx but we have 2 dx. Therefore, we need to change 2 dx to 3 dx. We do this
by writing

$$\int \frac{2}{3x + 1} \, dx = \left(\frac{3}{2}\right)\left(\frac{2}{3}\right) \int \frac{2}{3x + 1} \, dx$$

$$= \frac{2}{3} \int \frac{3}{2} \frac{2}{3x + 1} \, dx$$

$$= \frac{2}{3} \int \frac{3}{3x + 1} \, dx$$

$$= \frac{2}{3} \ln (3x + 1) + C$$

where the 2/3 is used to compensate for the 3/2
used in the integral.

Therefore

$$\int \frac{2}{3x + 1} \, dx = \frac{2}{3} \ln(3x + 1) + C$$

Practice Problems

Evaluate the following integrals:

1. $\int \dfrac{dx}{3x + 2}$

2. $\int \dfrac{dx}{5 - 2x}$

3. $\int \dfrac{x}{2 - 3x^2} \, dx$

4. $\int \dfrac{2x^3}{3 + 2x^4} \, dx$

Answers

1. $\dfrac{1}{3} \ln(3x + 2) + C$

2. $-\dfrac{1}{2} \ln(5 - 2x) + C$

3. $-\dfrac{1}{6} \ln(2 - 3x^2) + C$

4. $\dfrac{1}{4} \ln(3 + 2x^4) + C$

Method 3

In the third method that we will discuss, for solving integrals of quotients, we find that to integrate an algebraic function which has a numerator which is not of lower degree than the denominator we proceed as follows.

Change the integrand into a polynomial plus a fraction by dividing the denominator into the numerator. After this is accomplished, apply the rules available.

EXAMPLE: Evaluate

$$\int \frac{16x^2 - 4x - 8}{2x + 1} \, dx$$

SOLUTION: Divide the denominator into the numerator, then

$$\int \frac{16x^2 - 4x - 8}{2x + 1} \, dx = \int \left(8x - 6 - \frac{2}{2x + 1}\right) dx$$

19

$$= \int 8x \, dx - \int 6 \, dx - \int \frac{2}{2x+1} \, dx$$

and, integrating each separately, we have

$$\int 8x \, dx = 4x^2 + C_1$$

and

$$- \int 6 \, dx = -6x + C_2$$

and

$$- \int \frac{2}{2x+1} \, dx = -\ln(2x+1) + C_3$$

Then, by substitution, find that

$$\int \frac{16x^2 - 4x - 8}{2x+1} \, dx = 4x^2 - 6x - \ln(2x+1) + C$$

where

$$C = C_1 + C_2 + C_3$$

EXAMPLE: Evaluate

$$\int \frac{x}{x+1} \, dx$$

SOLUTION: The numerator is not of lower degree than the denominator; therefore we divide and find that

$$\int \frac{x}{x+1} \, dx = \int 1 - \frac{1}{x+1} \, dx$$

$$= \int dx - \int \frac{1}{x+1} \, dx$$

Integrating separately,

$$\int dx = x + C_1$$

and

$$-\int \frac{1}{x + 1} \, dx = -\ln(x + 1) + C_2$$

therefore

$$\int \frac{x}{x + 1} \, dx = x - \ln(x + 1) + C$$

where

$$C = C_1 + C_2$$

Practice Problems

Evaluate the following integrals:

1. $\int \frac{2x^2 + 6x + 5}{x + 1} \, dx$

2. $\int \frac{3x - 8}{x} \, dx$

3. $\int \frac{6x^3 + 13x^2 + 20x + 23}{2x + 3} \, dx$

4. $\int \frac{21x + 16x + 4}{3x + 1} \, dx$

Answers

1. $x^2 + 4x + \ln(x + 1) + C$

2. $3x - 8\ln(x) + C$

3. $x^3 + x^2 + 7x + \ln(2x + 3) + C$

4. $\frac{7}{2} x^2 + 3x + \frac{1}{3} \ln(3x + 1) + C$

Constant to a Variable Power

In this section a discussion of two forms of a constant to a variable power is presented. The two forms are a^u and e^u where u is the variable and a and e are the constants.

Formula:

$$\int a^u \, du = \frac{a^u}{\ln a} + C$$

Proof:

$$d(a^u + C_1) = a^u \ln a \, du$$

then

$$\int a^u \ln a \, du = a^u + C_1$$

but ln a is a constant, then

$$\int a^u \ln a \, du = \ln a \int a^u \, du$$

and

$$\ln a \int a^u \, du = a^u + C_1$$

Then, by dividing both sides by ln a, we have

$$\frac{\ln a}{\ln a} \int a^u \, du = \frac{a^u}{\ln a} + \frac{C_1}{\ln a}$$

and letting

$$C = \frac{C_1}{\ln a}$$

we have

$$\int a^u \, du = \frac{a^u}{\ln a} + C$$

EXAMPLE: Evaluate

$$\int 3^x \, dx$$

SOLUTION: Let

$$u = x$$

and

$$du = 1 \, dx$$

therefore, by knowing that

$$\int a^u \, du = \frac{a^u}{\ln a} + C$$

and using substitution, we find that

$$\int 3^x \, dx = \frac{3^x}{\ln 3} + C$$

EXAMPLE: Evaluate

$$\int 3^{2x} \, dx$$

SOLUTION: Let

$$u = 2x$$

and

$$du = 2\ dx$$

The integral should contain a factor of 2 in order that

$$du = 2dx$$

Thus we add a factor of 2 in the integral and compensate by multiplying the integral by 1/2.

Then

$$\int 3^{2x}\ dx = \frac{1}{2} \int (2)\ 3^{2x}\ dx$$

$$= \frac{1}{2} \int 3^{2x}\ 2\ dx$$

therefore

$$\frac{1}{2} \int 3^{2x}\ 2\ dx = (\frac{1}{2})\ \frac{3^{2x}}{\ln 3} + C$$

$$= \frac{3^{2x}}{2 \ln 3} + C$$

EXAMPLE: Evaluate

$$\int 7xb^{x^2}\ dx$$

24

SOLUTION: Let

$$u = x^2$$

and

$$du = 2x \, dx$$

In order to use

$$\int a^u \, du = \frac{a^u}{\ln a} + C$$

the integral must be in the form of

$$\int b^{x^2} \, 2x \, dx$$

but we have

$$\int b^{x^2} \, 7x \, dx$$

therefore we remove the 7 and insert a 2 by writing

$$\int 7xb^{x^2} \, dx = \int \left(\frac{7}{2}\right)\left(\frac{2}{7}\right) \, 7xb^{x^2} \, dx$$

$$= \frac{7}{2} \int \frac{2}{7} \, 7xb^{x^2} \, dx$$

$$= \frac{7}{2} \int 2xb^{x^2} \, dx$$

$$= \frac{7}{2} \int b^{x^2} \, 2x \, dx$$

25

$$= \frac{7}{2} \frac{b^{x^2}}{\ln b} + C$$

Practice Problems

Evaluate the following integrals:

1. $\int 10^{2x} \, dx$

2. $\int 7^{3x} \, dx$

3. $\int 9^{x^2} x \, dx$

4. $\int 2^{(3x^2 + 1)} x \, dx$

Answers

1. $\dfrac{10^{2x}}{2 \ln 10} + C$

2. $\dfrac{7^{3x}}{3 \ln 7} + C$

3. $\dfrac{9^{x^2}}{2 \ln 9} + C$

4. $\dfrac{2^{(3x^2 + 1)}}{6 \ln 2} + C$

We will now discuss the second form of the integral of a constant to a variable power.

Formula:

$$\int e^u \, du = e^u + C$$

Proof:

$$d(e^u + C) = e^u \, du$$

therefore

$$\int e^u \, du = e^u + C$$

EXAMPLE: Evaluate

$$\int e^x \, dx$$

SOLUTION: Let

$$u = x$$

and

$$du = 1 \, dx$$

The integral is in the correct form to use;

$$\int e^u \, du = e^u + C$$

therefore, using substitution, we find

$$\int e^x \, dx = e^x + C$$

EXAMPLE: Evaluate

$$\int e^{2x} \, dx$$

SOLUTION: Let

$$u = 2x$$

and

$$du = 2 \ dx$$

We need a factor of 2 in the integral and write

$$\int e^{2x} \ dx = \frac{2}{2} \int e^{2x} \ dx$$

$$= \frac{1}{2} \int e^{2x} \ 2dx$$

$$= \frac{1}{2} e^{2x} + C$$

EXAMPLE: Evaluate

$$\int xe^{2x^2} \ dx$$

SOLUTION: Let

$$u = 2x^2$$

and

$$du = 4x \ dx$$

Here a factor of 4 is needed in the integral, therefore

$$\int xe^{2x^2} \ dx = \int \frac{4}{4} \ xe^{2x^2} \ dx$$

$$= \frac{1}{4} \int 4xe^{2x^2} \ dx$$

$$= \frac{1}{4} e^{2x^2} + C$$

EXAMPLE: Evaluate

$$\int \frac{x^2}{e^{x^3}} \, dx$$

SOLUTION: Write the integral

$$\int \frac{x^2}{e^{x^3}} \, dx = \int x^2 e^{-x^3} \, dx$$

and let

$$u = -x^3$$

and

$$du = -3x^2 \, dx$$

therefore

$$\int x^2 e^{-x^3} \, dx = -\frac{1}{3} \int -3x^2 e^{-x^3} \, dx$$

$$= -\frac{1}{3} e^{-x^3} + C$$

Practice Problems

Evaluate the following integrals:

1. $\int -2xe^{-x^2} \, dx$

2. $\int e^{4x} \, dx$

29

3. $\int e^{(2x - 1)}\ dx$

4. $\int \dfrac{2x}{e^{x^2}}\ dx$

Answers

1. $e^{-x^2} + C$

2. $\dfrac{1}{4} e^{4x} + C$

3. $\dfrac{1}{2} e^{(2x - 1)} + C$

4. $-e^{-x^2} + C$

Trigonometric Functions

Trigonometric functions, which comprise one group of transcendental functions, may be differentiated and integrated in the same fashion as the other functions. We will limit our proofs to the sine, cosine, and secant functions, but will list several others.

Formula:

$$\int \sin u\ du = -\cos u + C$$

Proof:

$$d(\cos u + C) = -\sin u\ du$$

and

$$d(-\cos u + C) = \sin u \, du$$

therefore

$$\int \sin u \, du = -\cos u + C$$

Formula:

$$\int \cos u \, du = \sin u + C$$

Proof:

$$d(\sin u + C) = \cos u \, du$$

therefore

$$\int \cos u \, du = \sin u + C$$

Formula:

$$\int \sec^2 u \, du = \tan u + C$$

Proof:

$$d(\tan u + C) = d\left(\frac{\sin u}{\cos u} + C\right)$$

and by the quotient rule

$$d\left(\frac{\sin u}{\cos u}\right) + C = \frac{\cos u(\cos u) - \sin u(-\sin u)}{\cos^2 u} \, du$$

$$= \left(\frac{\cos^2 u + \sin^2 u}{\cos^2 u}\right) du$$

$$= \left(\frac{1}{\cos^2 u}\right) du$$

31

$$= \sec^2 u \; du$$

therefore

$$\int \sec^2 u \; du = \tan u + C$$

To this point we have considered integrals of trigonometric functions which result in functions of the sine, cosine, and tangent. Those integrals which result in functions of the cotangent, secant, and cosecant are included in the following list of elementary integrals.

$$\int \sin u \; du = -\cos u + C$$

$$\int \cos u \; du = \sin u + C$$

$$\int \sec^2 u \; du = \tan u + C$$

$$\int \csc^2 u \; du = -\cot u + C$$

$$\int \sec u \tan u \; du = \sec u + C$$

$$\int \csc u \cot u \; du = -\csc u + C$$

EXAMPLE: Evaluate

$$\int \sin 3x \; dx$$

SOLUTION: We need the integral in the form of

$$\int \sin u \, du = -\cos u + C$$

therefore, we let

$$u = 3x$$

and

$$du = 3 \, dx$$

but we do not have 3 dx. Therefore, we multiply the integral by 3/3 and rearrange as follows:

$$\int \sin 3x \, dx = \frac{3}{3} \int \sin 3x \, dx$$

$$= \frac{1}{3} \int \sin 3x \, 3 \, dx$$

then

$$\frac{1}{3} \int \sin 3x \, 3 \, dx = \frac{1}{3} (-\cos 3x) + C$$

$$= -\frac{1}{3} \cos 3x + C$$

EXAMPLE: Evaluate

$$\int \cos(2x + 4) \, dx$$

SOLUTION: Let

$$u = (2x + 4)$$

and

$$du = 2 \, dx$$

33

Therefore,

$$\int \cos(2x + 4)\ dx = \frac{2}{2} \int \cos(2x + 4)\ dx$$

$$= \frac{1}{2} \int \cos(2x + 4)\ 2\ dx$$

$$= \frac{1}{2} \sin(2x + 4) + C$$

EXAMPLE: Evaluate

$$\int (3 \sin 2x + 4 \cos 3x)\ dx$$

SOLUTION: We use the rule for sums and write

$$\int (3 \sin 2x + 4 \cos 3x)\ dx$$
$$= \int 3 \sin 2x\ dx + \int 4 \cos 3x\ dx$$

Then, in the integral

$$\int 3 \sin 2x\ dx$$

let

$$u = 2x$$

and

$$du = 2\ dx$$

but we have

$$3\ dx$$

To change 3 dx to 2 dx we divide by 3 and multiply by 2, with proper compensation, as follows:

$$\int 3 \sin 2x \, dx = \left(\frac{2}{3}\right)\left(\frac{3}{2}\right) \int 3 \sin 2x \, dx$$

$$= \frac{3}{2} \int \frac{2}{3} (3 \sin 2x \, dx)$$

$$= \frac{3}{2} \int 2 \sin 2x \, dx$$

$$= \frac{3}{2} (-\cos 2x) + C_1$$

$$= -\frac{3}{2} \cos 2x + C_1$$

The second integral

$$\int 4 \cos 3x \, dx$$

with

$$u = 3x$$

and

$$du = 3 \, dx$$

is evaluated as follows:

$$\int 4 \cos 3x \, dx = \left(\frac{4}{3}\right)\left(\frac{3}{4}\right) \int 4 \cos 3x \, dx$$

$$= \frac{4}{3} \int 3 \cos 3x \, dx$$

$$= \frac{4}{3} (\sin 3x) + C_2$$

Then, by combining the two solutions, we have

$$\int (3 \sin 2x + 4 \cos 3x) \, dx$$

$$= -\frac{3}{2} \cos 2x + C_1 + \frac{4}{3} \sin 3x + C_2$$

$$= -\frac{3}{2} \cos 2x + \frac{4}{3} \sin 3x + C$$

where

$$C_1 + C_2 = C$$

EXAMPLE: Evaluate

$$\int \sec^2 3x \, dx$$

SOLUTION: Let

$$u = 3x$$

and

$$du = 3 \, dx$$

We need 3 dx so we write

$$\int \sec^2 3x \, dx = \frac{3}{3} \int \sec^2 3x \, dx$$

$$= \frac{1}{3} \int \sec^2 3x \, 3 \, dx$$

$$= \frac{1}{3} (\tan 3x) + C$$

EXAMPLE: Evaluate

$$\int \csc 2x \cot 2x \, dx$$

SOLUTION: Let

$$u = 2x$$

and

$$du = 2 \, dx$$

We require du equal to 2 dx so we write

$$\int \csc 2x \cot 2x \, dx = \frac{2}{2} \int \csc 2x \cot 2x \, dx$$

$$= \frac{1}{2} \int 2 \csc 2x \cot 2x \, dx$$

$$= -\frac{1}{2} \csc 2x + C$$

EXAMPLE: Evaluate

$$\int \csc^2 3x \, dx$$

SOLUTION: Let

$$u = 3x$$

and

$$du = 3 \, dx$$

37

then

$$\int \csc^2 3x \, dx = \frac{1}{3} \int 3 \csc^2 3x \, dx$$

$$= -\frac{1}{3} \cot 3x + C$$

EXAMPLE: Evaluate

$$\int \sec \frac{x}{2} \tan \frac{x}{2} \, dx$$

SOLUTION: Let

$$u = \frac{x}{2}$$

and

$$du = \frac{1}{2} \, dx$$

then

$$\int \sec \frac{x}{2} \tan \frac{x}{2} \, dx = 2 \int \frac{1}{2} \sec \frac{x}{2} \tan \frac{x}{2} \, dx$$

$$= 2 \sec \frac{x}{2} + C$$

Practice Problems

Evaluate the following integrals:

1. $\int \cos 4x \, dx$

2. $\int \sin 5x \, dx$

3. $\int \sec^2 6x \, dx$

4. $\int 3 \cos(6x + 2) \, dx$

5. $\int x \sin(2x^2) \, dx$

6. $\int 2 \csc^2 5x \, dx$

7. $\int 3 \sec \frac{x}{3} \tan \frac{x}{3} \, dx$

Answers

1. $\frac{1}{4} \sin 4x + C$

2. $-\frac{1}{5} \cos 5x + C$

3. $\frac{1}{6} \tan 6x + C$

4. $\frac{1}{2} \sin(6x + 2) + C$

5. $-\frac{1}{4} \cos(2x^2) + C$

6. $-\frac{2}{5} \cot 5x + C$

7. $9 \sec \frac{x}{3} + C$

Trigonometric Functions of the Form $\int u^n \, du$

The integrals of powers of trigonometric functions will be limited to those which may, by substitution, be written in the form

$$\int u^n \, du$$

EXAMPLE: Evaluate

$$\int \sin^4 x \cos x \, dx$$

SOLUTION: Let

$$u = \sin x$$

and

$$du = \cos x \, dx$$

By substitution

$$\int \sin^4 x \cos x \, dx = \int u^4 \, du$$

$$= \frac{u^5}{5} + C$$

Then, by substitution again, find that

$$\frac{u^5}{5} + C = \frac{\sin^5 x}{5} + C$$

therefore

$$\int \sin^4 x \cos x \, dx = \frac{\sin^5 x}{5} + C$$

EXAMPLE: Evaluate

$$\int \cos^3 x(-\sin x) \, dx$$

SOLUTION: Let

$$u = \cos x$$

and

$$du = -\sin x \, dx$$

Write

$$\int u^3 \, du = \frac{u^4}{4} + C$$

and by substitution

$$\frac{u^4}{4} + C = \frac{\cos^4 x}{4} + C$$

Practice Problems

Evaluate the following integrals:

1. $\int \sin^2 x \cos x \, dx$

2. $\int \sin^4 x \cos x \, dx$

3. $\int 2 \sin x \cos x \, dx$

4. $\int \dfrac{\cos 2x}{\sin^3 2x} \, dx$

5. $\int \cos^3 x \sin x \, dx$

6. $\int \sin x \cos x \, (\sin x + \cos x) \, dx$

Answers

1. $\dfrac{1}{3} \sin^3 x + C$

2. $\dfrac{1}{5} \sin^5 x + C$

3. $\sin^2 x + C$

4. $\dfrac{-1}{4 \sin^2 2x} + C$

5. $-\dfrac{1}{4} \cos^4 x + C$

6. $\dfrac{\sin^3 x - \cos^3 x}{3} + C$

Chapter 16
Combinations & Permutations
for Probability & Statistics

This chapter deals with concepts required for the study of probability and statistics. Statistics is a branch of science which is an outgrowth of the theory of probability. Combinations and permutations are used in both statistics and probability, and they, in turn, involve operations with factorial notation. Therefore, combinations, permutations, and factorial notation are discussed in this chapter.

Definitions

A combination is defined as a possible selection of a certain number of objects taken from a group with no regard given to order. For instance, suppose we were to choose two letters from a group of three letters. If the group of three letters were A, B, and C, we could choose the letters in combinations of two as follows:

AB, AC, BC

The order in which we wrote the letters is of no concern. That is, AB could be written BA but we would still have only one combination of the letters A and B.

If order were considered, we would refer to the letters as permutations and make a distinction between AB and BA. The permutations of two letters from the group of three letters would be as follows:

AB, AC, BC, BA, CA, CB

The symbol used to indicate the foregoing combination will be $_3C_2$, meaning a group of three objects taken two at a time. For the previous permutation we will use $_3P_2$, meaning a group of three objects taken two at a time and ordered.

An understanding of factorial notation is required prior to a detailed discussion of combinations and permutations. We define the product of the integers 1 through n as n fractorial and use the symbol n! to denote this. That is,

$$3! = 1 \cdot 2 \cdot 3$$

$$6! = 1 \cdot 2 \cdot 3 \cdot 4 \cdot 5 \cdot 6$$

$$n! = 1 \cdot 2 \cdot 3 \cdots (n - 1) \cdot n$$

EXAMPLE: Find the value of 5!
SOLUTION: Write

$$5! = 1 \cdot 2 \cdot 3 \cdot 4 \cdot 5$$

$$= 120$$

EXAMPLE: Find the value of

$$\frac{5!}{3!}$$

SOLUTION: Write

$$5! = 5 \cdot 4 \cdot 3 \cdot 2 \cdot 1$$

and

$$3! = 3 \cdot 2 \cdot 1$$

then

$$\frac{5!}{3!} = \frac{5 \cdot 4 \cdot 3 \cdot 2 \cdot 1}{3 \cdot 2 \cdot 1}$$

and by simplification

$$\frac{5 \cdot 4 \cdot 3 \cdot 2 \cdot 1}{3 \cdot 2 \cdot 1} = 5 \cdot 4$$

$$= 20$$

The previous example could have been solved by writing

$$\frac{5!}{3!} = \frac{3! \; 4 \cdot 5}{3!}$$

$$= 5 \cdot 4$$

Notice that we wrote

$$5! = 5 \cdot 4 \cdot 3 \cdot 2 \cdot 1$$

45

and combined the factors

$$3 \cdot 2 \cdot 1$$

as

$$3!$$

then

$$5! = 3! \, 4 \cdot 5$$

EXAMPLE: Find the value of

$$\frac{6! - 4!}{4!}$$

SOLUTION: Write

$$6! = 4! \, 5 \cdot 6$$

and

$$4! = 4! \, 1$$

then

$$\frac{6! - 4!}{4!} = \frac{4! \, (5 \cdot 6 - 1)}{4!}$$

$$= (5 \cdot 6 - 1)$$

$$= 29$$

Notice that 4! was factored from the expression

$$6! - 4!$$

Theorem

If n and r are positive integers, with n greater than r, then

$$n! = r! \, (r + 1) \, (r + 2) \cdots n$$

This theorem allows us to simplify an expression as follows:

$$5! = 4! \, 5$$

$$= 3! \, 4 \cdot 5$$

$$= 2! \, 3 \cdot 4 \cdot 5$$

$$= 1 \cdot 2 \cdot 3 \cdot 4 \cdot 5$$

Another example is

$$(n + 2)! = (n + 1)! \, (n + 2)$$

$$= n! \, (n + 1) \, (n + 2)$$

$$= (n - 1)! \, n(n + 1) \, (n + 2)$$

EXAMPLE: Simplify

$$\frac{(n + 3)!}{n!}$$

SOLUTION: Write

$$(n + 3)! = n! \ (n + 1)(n + 2)(n + 3)$$

then

$$\frac{(n + 3)!}{n!} = \frac{n! \ (n + 1) \ (n + 2) \ (n + 3)}{n!}$$

$$= (n + 1) \ (n + 2) \ (n + 3)$$

Practice Problems

Find the value of problems 1-4 and simplify problems 5 and 6.

1. $6!$

2. $3! \ 4!$

3. $\dfrac{8!}{11!}$

4. $\dfrac{5! - 3!}{3!}$

5. $\dfrac{n!}{(n - 1)!}$

6. $\dfrac{(n + 2)!}{n!}$

Answers

1. 720
2. 144

3. $\dfrac{1}{990}$
4. 19
5. n
6. $(n + 1) (n + 2)$

Combinations

As indicated previously, a combination is the selection of a certain number of objects taken from a group of objects without regard to order. We use the symbol $_5C_3$ to indicate that we have five objects taken three at a time, without regard to order. Using the letters A, B, C, D, and E, to designate the five objects, we list the combinations as follows:

ABC   ABD   ABE   ACD   ACE

ADE   BCD   BCE   BDE   CDE

We find there are ten combinations of five objects taken three at a time. We made the selection of three objects, as shown, but we called these selections combinations. The word combinations infers that order is not considered.

EXAMPLE: Suppose we wish to know how many color combinations can be made from four different colored marbles, if we use only three marbles at a time. The marbles are colored red, green, white, and yellow.

SOLUTION: We let the first letter in each word indicate the color, then we list the possible combinations as follows:

RGW   RGY   RWY   GWY

If we rearrange the first group, RGW, to form GWR or RWG we still have the same color combination; therefore order is not important.

The previous examples are completely within our capabilities, but suppose we have 20 boys and wish to know how many different basketball teams we could form, one at a time, from these boys. Our listing would be quite lengthy and we would have a difficult task to determine that we had all of the possible combinations. In fact, there would be over 15,000 combinations we would have to list. This indicates the need for a formula for combinations.

Formula

The general formula for possible combinations of r objects from a group of n objects is

$$_nC_r = \frac{n(n-1)\ldots(n-r+1)}{1\cdot 2\cdot 3\cdots r}$$

The denominator may be written as

$$1\cdot 2\cdot 3\ldots r = r!$$

and if we multiply both numerator and denominator by

$$(n-r)\cdots 2\cdot 1$$

which is

$$(n-r)!$$

we have

$$_nC_r = \frac{n(n-1)\cdots(n-r+1)(n-r)\cdots2\cdot1}{r!\,(n-r)\cdots2\cdot1}$$

The numerator

$$n(n-1)\cdots(n-r+1)(n-r)\cdots2\cdot1$$

is

$$n!$$

Then

$$_nC_r = \frac{n!}{r!\,(n-r)!}$$

This formula is read: The number of combinations of n objects taken r at a time is equal to n factorial divided by r factorial times n minus r factorial.

EXAMPLE: In the previous problem where 20 boys were available, how many different basketball teams could be formed?

SOLUTION: If the choice of which boy played center, guard, or forward is not considered, we find we desire the number of combinations of 20 boys taken five at a time and write

$$_nC_r = \frac{n!}{r!\,(n-r)!}$$

where

$$n = 20$$

and

51

$$r = 5$$

Then, by substitution we have

$$_nC_r = {_{20}C_5} = \frac{20!}{5!\,(20-5)!}$$

$$= \frac{20!}{5!\,15!}$$

$$= \frac{15!\,16 \cdot 17 \cdot 18 \cdot 19 \cdot 20}{15!\,5!}$$

$$= \frac{16 \cdot 17 \cdot 18 \cdot 19 \cdot 20}{5 \cdot 4 \cdot 3 \cdot 2 \cdot 1}$$

$$= \frac{16 \cdot 17 \cdot 3 \cdot 19 \cdot 1}{1}$$

$$= 15,504$$

EXAMPLE: A man has, in his pocket, a silver dollar, a half-dollar, a quarter, a dime, a nickel, and a penny. If he reaches into his pocket and pulls out three coins, how many different sums may he have?

SOLUTION: The order in not important, therefore the number of combinations of coins possible is

$$_6C_3 = \frac{6!}{3!\,(6-3)!}$$

$$= \frac{6!}{3!\,3!}$$

$$= \frac{3!\,4 \cdot 5 \cdot 6}{3!\,3!}$$

$$= \frac{4 \cdot 5 \cdot 6}{3 \cdot 2 \cdot 1}$$

$$= \frac{4 \cdot 5}{1}$$

$$= 20$$

EXAMPLE: Find the value of

$$_3C_3$$

SOLUTION: We use the formula given and find that

$$_3C_3 = \frac{3!}{3! \, (3 - 3)!}$$

$$= \frac{3!}{3! \, 0!}$$

This seems to violate the rule, "division by zero is not allowed," but we define 0! as equal 1. Then

$$\frac{3!}{3! \, 0!} = \frac{3!}{3!} = 1$$

which is obvious if we list the combinations of three things taken three at a time.

Practice Problems

Find the value of problems 1-6 and simplify problems 7,8 and 9.

1. $_6C_2$

2. $_6C_4$

3. $_{15}C_5$

4. $_7C_7$

5. $\dfrac{_6C_3 + _7C_3}{_{13}C_6}$

6. $\dfrac{(_7C_3)\,(_6C_3)}{_{14}C_4}$

7. We want to paint three rooms in a house, each a different color and we may choose from seven different colors of paint. How many color combinations are possible, for the three rooms?

8. If 20 boys go out for the football team, how many different teams may be formed, one at a time?

9. Two boys and their dates go to the drive-in and each wants a different flavor ice cream cone. The drive-in has 24 flavors of ice cream. How many combinations of flavors may they choose?

Answers

1. 15

2. 15

3. 3,003

4. 1

5. $\dfrac{5}{156}$

6. $\dfrac{100}{143}$

7. 35

8. 167,960

9. 10,626

Additional Solved Problems

How many sums of money can be obtained by choosing two coins from a box containing a penny, a nickel, a dime, a quarter, and a half dollar?

SOLUTION:

The order makes no difference here, since a selection of a penny and a dime is the same as a selection of a dime and a penny insofar as the sum is concerned. This is a case of combinations, then, rather than permutations. The number of combinations of n different objects taken r at a time is equal to

$$\frac{n(n-1) \ldots (n-r+1)}{1 \times 2 \ldots r} \, .$$

In this example, $n = 5$, $r = 2$, therefore,

$$C\,(5,\,2) = \frac{5 \times 4}{1 \times 2} = 10 \, .$$

> How many baseball teams of nine members can be chosen from among 12 boys, without regard to the position played by each member?

SOLUTION:

Since there is no regard to position, this is a combinations problem (if order or arrangement had been important, it would have been a permutations problem). The general formula for the number of combinations of n items taken r at a time is

$$C\ (n,\ r) = \frac{n!}{r!\ (n-r)!}\ .$$

We have to find the number of combinations of 12 items taken nine at a time. Hence, we have

$$C\ (12,\ 9) = \frac{12!}{9!\ (12-9)!} = \frac{12!}{9!\ 3!} = \frac{12 \times 11 \times 10 \times 9!}{3 \times 2 \times 1 \times 9!} = 220\ .$$

Therefore, there are 220 possible teams.

> How many words, each consisting of two vowels and three consonants, can be formed from the letters in the word "integral"?

SOLUTION:

To find the number of ways to choose vowels or consonants from letters, we use combinations. The number of combinations of n different objects taken r at a time is defined to be

$$C\ (n,\ r) = \frac{n!}{r!\ (n-r)!}\ .$$

We first select the two vowels to be used from among the three vowels in "integral"; this can be done in $C\ (3,\ 2) = 3$ ways. Next, we select the three consonants from the five in "integral"; this yields $C\ (5,\ 3) = 10$ possible choices. To find the number of ordered arrangements of five letters selected five at a time, we need to find the number of permutations of choosing r from n objects. Symbolically, it is $P\ \{n,\ r\}$, which is defined to be

$$P\ \{n,\ r\} = \frac{n!}{(n-r)!}\ .$$

We permute the five chosen letters in all possible ways, of which there are $P\ \{5,\ 5\} = 5! = 120$ arrangements since no two letters are the same. Finally, to find the total number of words which can be formed, we apply the Fundamental Principle of Counting, which states that if one event can be performed in m ways, another one in n ways, and another in k ways, then the total number of ways in which all events can occur is $m \times n \times k$. Hence, the total number of possible words is, by the Fundamental Principle,

$$C\ (3,\ 2)\ C\ (5,\ 3)\ P\ \{5,\ 5\} = 3 \times 10 \times 120 = 3,600.$$

> How many different bridge hands are there?

SOLUTION:

A bridge hand contains 13 cards dealt from a 52-card deck. The order in which the cards are dealt is not important. The number of hands that might be dealt is the same as the number of hands it is possible to select, if one were allowed to select 13 cards at random from a standard deck. The question now becomes: how many ways may 52 objects be taken in combinations of 13 at a time? Let us denote this number by $_{52}C_{13}$.

The solution is $\binom{52}{13} = \dfrac{52!}{13!\ 39!}$.

With the help of tables for $n!$, we find the number of possible bridge hands to be about 635,000,000,000.

Principle of Choice

The principle of choice is discussed in relation to combinations although it is also, later in this chapter, discussed in relation to permutations. It is stated as follows:

If a selection can be made in n_1 ways, and after this selection is made, a second selection can be made in n_2 ways, and after this selection is made, a third selection can be made in n_3 ways, and so forth for r ways, then the r selections can be made together in

$$n_1 \cdot n_2 \cdot n_3 \cdots n_r \text{ ways}$$

EXAMPLE: In how many ways can a coach choose first a football team and then a basketball team if 18 boys go out for either team?

SOLUTION: First let the coach choose a football team. That is

$$_{18}C_{11} = \frac{18!}{11!\ (18 - 11)!}$$

$$= \frac{18!}{11! \; 7!}$$

$$= \frac{11! \; 12 \cdot 13 \cdot 14 \cdot 15 \cdot 16 \cdot 17 \cdot 18}{11! \; 7 \cdot 6 \cdot 5 \cdot 4 \cdot 3 \cdot 2 \cdot 1}$$

$$= 31,824$$

The coach now must choose a basketball team from the remaining seven boys. That is

$$_7C_5 = \frac{7!}{5! \; (7 - 5)!}$$

$$= \frac{7!}{5! \; 2!}$$

$$= \frac{5! \; 6 \cdot 7}{5! \; 2!}$$

$$= \frac{6 \cdot 7}{2}$$

$$= 21$$

Then, together, the two teams can be chosen in

$$(31,824)(21) = 668,304 \text{ ways}$$

EXAMPLE: A man ordering dinner has a choice of one meat dish from four, a choice of three vegetables from seven, one salad from three, and one dessert from four. How many different menus are possible?

SOLUTION: The individual combinations are as follows:

$$\text{meat} \ldots \ldots \ldots _4C_1$$

$$vegetable \ldots \ldots \cdot _7C_4$$

$$salad \ldots \ldots \ldots \cdot _3C_1$$

$$dessert \ldots \ldots \cdot _4C_1$$

The value of

$$_4C_1 = \frac{4!}{1!\,(4-1)!}$$

$$= \frac{4!}{3!}$$

$$= 4$$

and

$$_7C_4 = \frac{7!}{4!\,(7-4)!}$$

$$= \frac{7!}{4!\,3!}$$

$$= \frac{5 \cdot 6 \cdot 7}{2 \cdot 3}$$

$$= 35$$

and

$$_3C_1 = \frac{3!}{1!\,(3-1)!}$$

$$= \frac{3!}{2!}$$

$$= 3$$

therefore, there are

$$(4) (35) (3) (4) = 1680$$

different menus available to the man.

Practice Problems

Solve the following problems.

1. A man has 12 different colored shirts and 20 different ties. How many shirt and tie combinations can he select to take on a trip, if he takes three shirts and five ties?

2. A petty officer, in charge of posting the watch, has in the duty section 12 men. He must post three different fire watches, then post four aircraft guards on different aircraft. How many different assignments of men can he make?

3. If there are 10 third class and 14 second class petty officers in a division which must furnish two second class and six third class petty officers for shore patrol, how many different shore patrol parties can be made?

Answers

1. 3,410,880
2. 27,720
3. 19,110

Permutations

Permutations are similar to combinations but extend the requirements of combinations by considering order.

Suppose we have two letters, A and B, and wish to know how many arrangements of these letters can be made. It is obvious that the answer is two. That is

AB and BA

If we extend this to the three letters A, B, and C, we find the answer to be

ABC, ACB, BAC, BCA, CAB, CBA

We had three choices for the first letter, and after we chose the first letter, we had only two choices for the second letter, and after the second letter, we had only one choice. This is shown in the "tree" diagram in figure 16-1. Notice that there is a total of six different paths to the ends of the "branches" of the "tree" diagram.

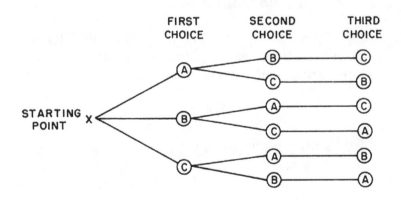

Figure 16-1.—"Tree" diagram.

If the number of objects is large, the tree would become very complicated; therefore, we approach the problem in another manner, using parentheses to show the possible choices. Suppose we were to arrange five objects in as many different orders as possible. We have for the first choice six objects.

$$(6) ( ) ( ) ( ) ( ) ( )$$

For the second choice we have only five choices.

$$(6) (5) ( ) ( ) ( ) ( )$$

For the third choice we have only four choices.

$$(6) (5) (4) ( ) ( ) ( )$$

This may be continued as follows:

$$(6) (5) (4) (3) (2) (1)$$

By applying the principle of choice we find the total possible ways of arranging the objects to be the product of the individual choices. That is

$$6 \cdot 5 \cdot 4 \cdot 3 \cdot 2 \cdot 1$$

and this may be written as

$$6!$$

This leads to the statement: The number of permutations of n objects, taken all together, is equal to n!.

EXAMPLE: How many permutations of seven different letters may be made?

SOLUTION: We could use the "tree" but this would become complicated. (Try it to see why.) We could use the parentheses as follows:

$$(7)\ (6)\ (5)\ (4)\ (3)\ (2)\ (1) = 5040$$

The easiest solution is to use the previous statement and write

$$_7P_7 = 7!$$

That is, the number of possible arrangements of seven objects, taken seven at a time, is $7!$. NOTE: Compare this with the number of COMBINATIONS of seven objects, taken seven at a time.

If we are faced with finding the number of permutations of seven objects taken three at a time, we use three parentheses as follows:
In the first position we have a choice of seven objects.

$$(7)\ (\ )\ (\ )$$

In the second position we have a choice of six objects

$$(7)\ (6)\ (\ )$$

In the last position we have a choice of five objects,

$$(7)\ (6)\ (5)$$

and by principle of choice, the solution is

$$7 \cdot 6 \cdot 5 = 210$$

Formula

At this point we will use our knowledge of combinations to develop a formula for the number of permutations of n objects taken r at a time.

Suppose we wish to find the number of permutations of five things taken three at a time. We first determine the number of groups of three, as follows:

$$_5C_3 = \frac{5!}{3! \, (5 - 3)!}$$

$$= \frac{5!}{3! \, 2!}$$

$$= 10$$

Thus, there are 10 groups of three objects each.

We are now asked to arrange each of these ten groups in as many orders as possible. We know that the number of permutations of three objects, taken together, is $3!$. We may arrange each of the 10 groups in $3!$ or six ways. The total number of possible permutations of $_5C_3$ then is

$$_5C_3 \cdot 3! = 10 \cdot 6$$

which is written

$$_5C_3 \cdot 3! = \, _5P_3$$

Put into the general form, then

$$n^Cr \cdot r! = n^Pr$$

and knowing that

$$n^Cr = \frac{n!}{r! \, (n - r)!}$$

then

$$n^Cr \cdot r! = \frac{n!}{r! \, (n - r)!} \cdot r!$$

$$= \frac{n!}{(n - r)!}$$

but

$$n^Cr \cdot r! = n^Pr$$

therefore

$$n^Pr = \frac{n!}{(n - r)!}$$

EXAMPLE: How many permutations of six objects, taken two at a time, can be made?

SOLUTION: The number of permutations of six objects, taken two at a time, is written

$$6^P2 = \frac{6!}{(6 - 2)!}$$

$$= \frac{6!}{4!}$$

$$= \frac{4! \; 5 \cdot 6}{4!}$$

$$= 5 \cdot 6$$

$$= 30$$

EXAMPLE: In how many ways can eight people be arranged in a row?

SOLUTION: All eight people must be in the row; therefore, we want the number of permutations of eight people, taken eight at a time, which is

$$_8P_8 = \frac{8!}{(8 - 8)!}$$

$$= \frac{8!}{0!}$$

(Remember that 0! was defined as equal to 1) then

$$\frac{8!}{0!} = \frac{8 \cdot 7 \cdot 6 \cdot 5 \cdot 4 \cdot 3 \cdot 2 \cdot 1}{1}$$

$$= 40,320$$

Problems dealing with combinations and permutations often require the use of both formulas to solve one problem.

EXAMPLE: There are eight first class and six second class petty officers on the board of the fifty-six club. In how many ways can they elect, from the board, a president, a vice-president, a secretary, and a treasurer if the president and secretary must be first class petty officers and the vice-president and treasurer must be second class petty officers?

SOLUTION: Since two of the eight first class petty officers are to fill two different offices, we write

$$_8P_2 = \frac{8!}{(8-2)!}$$

$$= \frac{8!}{6!}$$

$$= 7 \cdot 8$$

$$= 56$$

Then, two of the six second class petty officers are to fill two different offices; thus we write

$$_6P_2 = \frac{6!}{(6-2)!}$$

$$= \frac{6!}{4!}$$

$$= 5 \cdot 6$$

$$= 30$$

The principle of choice holds in this case; therefore, there are

$$56 \cdot 30 = 1680$$

ways to select the required office holders. The problem, thus far, is a permutation problem, but suppose we are asked the following: In how many ways can they elect the office holders from the board, if two of the office holders must be first class petty officers and two of the office holders must be second class petty officers?

SOLUTION: We have already determined how many ways eight things may be taken two at a time and how many ways six may be taken two at a time, and also, how many ways they may be taken together. That is

$$_8P_2 = 56$$

and

$$_6P_2 = 30$$

then

$$_8P_2 \cdot {}_6P_2 = 1680$$

Our problem now is to find how many ways we can combine the four offices, two at a time. Therefore, we write

$$_4C_2 = \frac{4!}{2!\,(4-2)!}$$

$$= \frac{4!}{2!\,2!}$$

$$= \frac{4\cdot3\cdot2\cdot1}{2\cdot2}$$

$$= 6$$

Then, in answer to the problem, we write

$$_8P_2 \cdot {}_6P_2 \cdot {}_4C_2 = 10,080$$

In words, if there are $_4C_2$ ways of combining the four offices, and then, for every one of these ways there are $_8P_2 \cdot {}_6P_2$ ways of arranging the office holders, then there are

$$_8P_2 \cdot {}_6P_2 \cdot {}_4C_2$$

ways of electing the petty officers.

Practice Problems

Find the answers to the following.

1. $_6P_3$

2. $_4P_3$

3. $_7P_2 \cdot _5P_2$

4. In how many ways can six people be seated in a row?

5. There are seven boys and nine girls in a club. In how many ways can they elect four different officers designated by A, B, C, and D if:

    (a) A and B must be boys and C and D must be girls?

    (b) two of the officers must be boys and two of the officers must be girls?

Answers

1. 120
2. 24
3. 840
4. 720
5. (a) 3,024
   (b) 18,144

In the question "How many different arrangements of the letters in the word STOP can be made?" were asked, we would write

$$_4P_4 = \frac{4!}{(4-4)!}$$

$$= \frac{4!}{0!}$$

$$= 24$$

We would be correct since all letters are different. If some of the letters were the same, we would reason as given in the following problem.

EXAMPLE: How many different arrangements of the letters in the word ROOM can be made?

SOLUTION: We have two letters alike. If we list the possible arrangements, using subscripts to make a distinction between the O's, we have

$R\ O_1 O_2 M$	$O_1 O_2 M\ R$	$O_1 M\ O_2 R$	$M\ O_1 O_2 R$
$R\ O_2 O_1 M$	$O_2 O_1 M\ R$	$O_2 M\ O_1 R$	$M\ O_2 O_1 R$
$R\ O_1 M\ O_2$	$O_1 O_2 R\ M$	$O_1 R\ M\ O_2$	$M\ O_1 R\ O_2$
$R\ O_2 M\ O_1$	$O_2 O_1 R\ M$	$O_2 R\ M\ O_1$	$M\ O_2 R\ O_1$
$R\ M\ O_1 O_2$	$O_1 M\ R\ O_2$	$O_1 R\ O_2 M$	$M\ R\ O_1 O_2$
$R\ M\ O_2 O_1$	$O_2 M\ R\ O_1$	$O_2 R\ O_1 M$	$M\ R\ O_2 O_1$

but we cannot distinguish between the O's and $R\ O_1 O_2 M$ and $R\ O_2 O_1 M$ would be the same arrangement without the subscript. Notice in the list that there are only half as many arrangements without the use of subscripts or a total of twelve arrangements. This leads to the statement: The number of arrangements of n items, where r_1, r_2, and r_k are alike, is given by

$$\frac{n!}{r_1! \; r_2! \; \cdots r_k!}$$

In the previous example n was equal to four and there were two letters alike; therefore, we would write

$$\frac{4!}{2!} = \frac{4 \cdot 3 \cdot 2 \cdot 1}{2 \cdot 1}$$

$$= 12$$

EXAMPLE: How many arrangements can be made using the letters in the word ADAPTATION?

SOLUTION: We use

$$\frac{n!}{r_1! \; r_2! \; \cdots r_k!}$$

where

$$n = 10$$

and

$$r_1 = 2 \text{ (two T's)}$$

and

$$r_2 = 3 \text{ (three A's)}$$

Then

$$\frac{n!}{r_1! \; r_2! \cdots r_k!} = \frac{10!}{2! \; 3!}$$

$$= \frac{2 \cdot 5 \cdot 6 \cdot 7 \cdot 8 \cdot 9 \cdot 10}{1}$$

$$= 302,400$$

Practice Problems

Find the number of possible arrangements of the letters in the following words.

1. DOWN
2. STRUCTURE
3. BOOK
4. MILLIAMPERE
5. TENNESSEE

Answers

1. 24
2. 45,360
3. 12
4. 2,494,800
5. 3,780

Although the previous discussions have been associated with formulas, problems dealing with combinations and permutations may be analyzed and solved in a more meaningful way without complete reliance upon the formulas.

EXAMPLE: How many four-digit numbers can be formed from the digits 2, 3, 4, 5, 6, and 7

    (a) without repetitions?

    (b) with repetitions?

SOLUTION: The (a) part of the question is a straight forward permutation problem and we reason that we want the number of permutations of six items taken four at a time.

Therefore

$$_6P_4 = \frac{6!}{(6-4)!}$$

$$= \frac{6 \cdot 5 \cdot 4 \cdot 3 \cdot 2 \cdot 1}{2 \cdot 1}$$

$$= 360$$

The (b) part of the question would become quite complicated if we tried to use the formulas; therefore, we reason as follows:

We desire a four digit number and find we have six choices for the first digit. That is, we may use any of the digits 2, 3, 4, 5, 6, or 7 in the thousands column which gives us six choices for the digit to be placed in the thousands column. If we select the digit 4 for the thousands column we still have a choice of any of the digits 2, 3, 4, 5, 6, or 7 for the hundreds column. This is because we are allowed repetition and may select the digit 4 for the hundreds column as we did for the thousands column. This gives us six choices for the hundreds column.

Continuing this reasoning, we could write the number of choices for each place value column as shown in table 16-1.

Table 16-1. — Place value choices.

thousands column	hundreds column	tens column	units column
six choices	six choices	six choices	six choices

In table 16-1, observe that the total number of choices for the four digit number, by the principle of choice, is

$$6 \cdot 6 \cdot 6 \cdot 6 = 1,296$$

Suppose, in the previous problem, we were to find how many three-digit odd numbers could be formed from the given digits, without repetition. We would be required to start in the units column because an odd number is determined by the units column digit. Therefore, we have only three choices. That is, either the 3, 5, or 7. For the tens column we have five choices and for the hundreds column we have four choices. This is shown in table 16-2.

Table 16-2. — Place value choices.

hundreds column	tens column	units column
four choices	five choices	three choices

In table 16-2, observe that there are

$$4 \cdot 5 \cdot 3 = 60$$

three-digit odd numbers that can be formed from the digits 2, 3, 4, 5, 6, and 7, without repetition.

Practice Problems

Solve the following problems.

1. Using the digits 4, 5, 6, and 7, how many two-digit numbers can be formed:
   (a) without repetitions?
   (b) with repetitions?

2. Using the digits 4, 5, 6, 7, 8, and 9, how many five-digit numbers can be formed:
   (a) without repetitions?
   (b) with repetitions?

3. Using the digits of problem 2, how many four-digit odd numbers can be formed, without repetitions?

Answers

1. (a) 12
   (b) 16
2. (a) 720
   (b) 7,776
3. 180

Additional Solved Problems

Calculate the number of permutations of the letters a, b, c, and d taken two at a time.

SOLUTION:

The first of the two letters may be taken in four ways (a, b, c, d). The second letter may therefore be selected from the remaining three letters in three ways. By the Fundamental Principle, the total number of ways of selecting two letters is equal to the product of the number of ways of selecting each letter, or

$4 \times 3 = 12$

The list of these permutations is:

ab	ba	ca	da
ac	bc	cb	db
ad	bd	cd	dc

Calculate the number of permutations of the letters a, b, c, and d taken four at a time.

SOLUTION:

The number of permutations of the four letters taken four at a time equals the number of ways the four letters can be arranged or ordered. Consider four places to be filled by the four letters. The first place can be filled in four ways, choosing from the four letters. The second place may be filled in three ways, selecting one of the three remaining letters. The third place may be filled in two ways with one of the two still remaining. The fourth place is filled one way with the last letter. By the Fundamental Principle, the total number of ways of ordering the letters equals the product of the number of ways of filling each ordered place, or $4 \times 3 \times 2 \times 1 = 24 = P(4, 4) = 4!$ (read "four factorial").

In general, for n objects taken r at a time,

$$P(n, r) = n(n-1)(n-2) \ldots (n-r+1) = \frac{r!}{(n-r)!} \quad (r < n).$$

For the special case where $r = n$,

$$P(n, n) = n(n-1)(n-2) \ldots (3)(2)(1) = n!,$$

since $(n-r)! = 0!$ which equals one by definition.

In how many different ways may three books be placed next to each other on a shelf?

SOLUTION:

We construct a pattern of three boxes to represent the places where the three books are to be placed next to each other on the shelf:

Since there are three books, the first place may be filled in three ways. There are then two books left, so that the second place may be filled in two ways. There is only one book left to fill the last place. Hence, our boxes take the following form:

3	2	1

The Fundamental Principle of Counting states that if one task can be performed in *a* different ways and, when it is performed in any one of these ways, a second task can be performed in *b* different ways, and a third task can be done in *c* ways, ... then all the tasks in succession can be performed in $a \times b \times c...$ different ways. Thus, the books can be arranged in $3 \times 2 \times 1 = 6$ ways. This can also be seen as follows. Since the arrangement of books on the shelf is important, this is a permutations problem. Recalling the general formula for the number of permutations of *n* items taken *r* at a time,

$$_nP_r = \frac{n!}{(n-r)!}, \text{ we replace } n \text{ by 3 and } r \text{ by 3 to obtain}$$

$$_3P_3 = \frac{3!}{(3-3)!} = \frac{3!}{0!} = \frac{3 \times 2 \times 1}{1} = 6.$$

Determine the number of distinct permutations of the letters in the word "banana."

SOLUTION:

In solving this problem we use the fact that the number of permutations *P* of *n* objects taken all at a time [P (n, n)], of which n_1 are alike, n_2 others are alike, n_3 others are alike, etc. is

$$P = \frac{n!}{n_1! \ n_2! \ n_3!...}, \quad \text{with } n_1 + n_2 + n_3 + ... = n .$$

In the given problem there are six letters ($n = 6$), of which two are alike, (there are two N's so that $n_1 = 2$), three others are alike (there are three A's, so that $n_2 = 3$), and one is left (there is one B, so $n_3 = 1$). Notice that $n_1 + n_2 + n_3 = 2 + 3 + 1 = 6 = n$; thus,

$$P = \frac{6!}{2! \ 3! \ 1!} = \frac{6 \times 5 \times 4 \times 3!}{2 \times 1 \times 3! \times 1} = 60 .$$

Thus, there are 60 permutations of the letters in the word "banana."

> Find the number of distinct permutations of the seven letters in the word "algebra."

SOLUTION:

A permutation is an ordered arrangement of a set of objects. For example, if you are given four letters a, b, c, and d and you choose two at a time, some permutations you can obtain are: ab, ac, ad, ba, bc, bd, ca, and cb.

For n items, we can arrange the first object in n different ways, the second in $n - 1$ different ways, the third in $n - 2$ different ways, etc. Thus, the n objects can be arranged in order in

$$n! = n \times (n - 1) \times (n - 2) \ldots 1 \text{ ways.}$$

Temporarily place subscripts, 1 and 2, on the a's to distinguish them, so that we now have $7! = 5,040$ possible permutations of the seven distinct objects. Of these 5,040 arrangements, half will contain the a's in the order a_1, a_2 and the other half will contain them in the order a_2, a_1. If we assume the two a's are indistinct, then we apply the following theorem. The number P of distinct permutations of n objects taken at a time, of which n_1 are alike, n_2 are alike of another kind, ... n_k are alike of still another kind, with

$n_1 + n_2 + \ldots + n_k = n$, is $P = \dfrac{n!}{n_1! \, n_2! \ldots n_k!}$. Then in this example, the two

a's are alike so $P = \dfrac{7!}{2!} = 2,520$ permutations of the letters in the word al-

gebra, when the a's are indistinguishable.

> In how many ways may a party of four women and four men be seated at a round table if the women and men are to occupy alternate seats?

SOLUTION:

If we consider the seats indistinguishable, then this is a problem in circular permutations, as opposed to linear permutations. In the standard linear permutation approach, each chair is distinguishable from the others. Thus, if a woman is seated first, she may be chosen four ways, then a man seated next to her may be chosen four ways, the next woman can be chosen three ways and the man next to her can be chosen three ways... Our diagram to the linear approach shows the number of ways each seat can be occupied.

By the Fundamental Principle of Counting there are thus $4 \times 4 \times 3 \times 3 \times 2 \times 2 \times 1 \times 1 = 576$ ways to seat the people.

However, if the seats are indistinguishable, then so long as each person has the same two people on each side, the seating arrangement is considered the same. Thus, we may suppose one person, say a woman, is seated in a particular place, and then arrange the remaining three women and four men relative to her. Because of the alternate seating scheme, there are three possible places for the remaining three women and four possible places for the four men. Hence, the total number of arrangements is $(3!) - (4!) = 6 \times 24 = 144$. In general, the formula for circular permutations of n items and n other items which are alternating is $(n - 1)! \, n!$. In our case, we have

$$(4 - 1)! \, 4! = 3! \, 4! = 3 \times 2 \times 4 \times 3 \times 2 = 144 .$$

Chapter 17
Probability

The history of probability theory dates back to the 17th century and at that time was related to games of chance. In the 18th century it was seen that probability theory had applications beyond the scope of games of chance. Some of the applications in which probability theory is applied are situations with outcomes such as life or death and boy or girl. In the present century, statistics and probability are applied to insurance, annuities, biology, and social investigations.

The treatment of probability in this chapter is limited to simple applications. These applications will be, to a large extent, based on games of chance which lend themselves to an understanding of basic ideas of probability.

Basic Concepts

If a coin were tossed, the chance that it would land heads up is just as likely as the chance it would land tails up. That is, the coin has no more reason to land heads up than it has to land tails up. Every toss of the coin is called a trial. We define probability as the ratio of the different number of ways a trial can succeed (or fail) to the total numbers of ways in which it may result. We will let p represent the probability of success and q represent the probability of failure.

One commonly misunderstood concept of probability is the effect prior trials have on a single trial. That is, after a coin has been tossed many times and every trial resulted in the coin falling heads up, will the next toss of the coin result in tails up? The answer is "not necessarily," and will be explained later in this chapter.

Probability of Success

If a trial must result in any of n equally likely ways, and if s is the number of successful ways and f is the number of failing ways, then the probability of success is

$$p = \frac{s}{s + f}$$

where

$$s + f = n$$

EXAMPLE: What is the probability that a coin will land heads up?

SOLUTION: There is only one way the coin can land heads up, therefore s equals one. There is also only one way the coin can land other than heads up; therefore, f equals one. Then

$$s = 1$$

and

$$f = 1$$

Thus the probability of success is

$$p = \frac{s}{s + f}$$

$$= \frac{1}{1 + 1}$$

$$= \frac{1}{2}$$

This, then, is the ratio of successful ways in which the trial can succeed to the total number of ways the trial can result.

EXAMPLE: What is the probability that a die (singular of dice) will land with a three showing on the upper face.

SOLUTION: There is only one favorable way the die may land and there are a total of five ways it can land without the three face up.

$$s = 1$$

and

$$f = 5$$

and

$$p = \frac{s}{s + f}$$

$$= \frac{1}{1 + 5}$$

$$= \frac{1}{6}$$

EXAMPLE: What is the probability of drawing a black marble from a box of marbles if all six of the marbles in the box are white?

SOLUTION: There are no favorable ways of success and there are six total ways, therefore,

$$s = 0$$

and

$$f = 6$$

then

$$p = \frac{0}{0 + 6}$$

$$= \frac{0}{6}$$

$$= 0$$

EXAMPLE: What is the probability of drawing a black marble from a box of six black marbles?

SOLUTION: There are six successful ways and no unsuccessful ways of drawing the marble, therefore

$$s = 6$$

and

$$f = 0$$

then

$$p = \frac{6}{6 + 0}$$

83

$$= \frac{6}{6}$$

$$= 1$$

The previous two examples are the extremes of probabilities and intuitively demonstrate that the probability of an event ranges from zero to one inclusive.

EXAMPLE: A box contains six hard lead pencils and twelve soft lead pencils. What is the probability of drawing a soft lead pencil from the box?

SOLUTION: We are given

$$s = 12$$

and

$$f = 6$$

then

$$p = \frac{12}{12 + 6}$$

$$= \frac{12}{18}$$

$$= \frac{2}{3}$$

Practice Problems

1. What is the probability of drawing an ace from a standard deck of fifty-two playing cards?

2. What is the probability of drawing a black ace from a standard deck of playing cards?

3. If a die is rolled, what is the probability of an odd number showing on the upper face?

4. A man has three nickels, two dimes, and four quarters in his pocket. If he draws a single coin from his pocket, what is the probability that:

    (a) he draws a nickel?

    (b) he draws a half-dollar?

    (c) he draws a quarter?

Answers

1. $\dfrac{1}{13}$

2. $\dfrac{1}{26}$

3. $\dfrac{1}{2}$

4. (a) $\dfrac{1}{3}$

   (b) 0

   (c) $\dfrac{4}{9}$

Probability of Failure

If a trial results in any of n equally likely ways, and s is the number of successful ways and f is the number of failures then, as before,

$$s + f = n$$

or

$$n - s = f$$

The probability of failure is given by

$$q = \frac{f}{s + f}$$

$$= \frac{n - s}{n}$$

A trial must result in either success or failure. If success is certain then p equals one and q equals zero. If success is impossible then p equals zero and q equals one. By combining both events—that is, in either case—the probability of success plus the probability of failure is equal to one as shown by

$$p = \frac{s}{s + f}$$

and

$$q = \frac{f}{s + f}$$

then

$$p + q = \frac{s}{s + f} + \frac{f}{s + f}$$

$$= 1$$

If, in any event

$$p + q = 1$$

then

$$q = 1 - p$$

In the case of tossing a coin, the probability of success is

$$p = \frac{s}{s + f}$$

$$= \frac{1}{1 + 1}$$

$$= \frac{1}{2}$$

and the probability of failure is

$$q = 1 - p$$

$$= 1 - \frac{1}{2}$$

$$= \frac{1}{2}$$

EXAMPLE: What is the probability of not drawing a black marble from a box containing six white, three red, and two black marbles from a box containing six white, three red, and two black marbles?

SOLUTION: The probability of drawing a black marble from the box is

$$p = \frac{s}{s + f}$$

$$= \frac{2}{2 + 9}$$

$$= \frac{2}{11}$$

Since the probability of drawing a marble is one, then the probability of not drawing a black marble is

$$q = 1 - p$$

$$= 1 - \frac{2}{11}$$

$$= \frac{9}{11}$$

Practice Problems

Compare the following problems and answers with the preceding problems dealing with the probability of success.

1. What is the probability of not drawing an ace from a standard deck of fifty-two playing cards?

2. What is the probability of not drawing a black ace from a standard deck of playing cards?

3. If a die is rolled, what is the probability of an odd number not showing on the upper face?

4. A man has three nickels, two dimes, and four quarters in his pocket. If he draws a single coin from his pocket, what is the probability that:

    (a) he does not draw a nickel?

    (b) he does not draw a half-dollar?

    (c) he does not draw a quarter?

Answers

1. $\dfrac{12}{13}$

2. $\dfrac{25}{26}$

3. $\dfrac{1}{2}$

4. (a) $\dfrac{2}{3}$

    (b) 1

    (c) $\dfrac{5}{9}$

Expectations

In this discussion of expectation we will consider two types. One is a numerical expectation and the other is value expectation.

Numerical Expectation

If you tossed a coin fifty times you would expect the coin to fall heads about twenty-five times. Your assumption is explained by the following definition.

If the probability of success in one trial is p, and k is the total number of trials, then pk is the expected number of successes in the k trials.

In the above example of tossing the coin fifty times the expected number of heads (successes) is

$$E_n = pk$$

where

E_n = expected number

p = probability of heads (successes)

k = number of tosses

Substituting values in the equation, we find that

$$E_n = (\frac{1}{2})\,50$$

$$= 25$$

EXAMPLE: A die is rolled by a player. What is the expectation of rolling a six in 30 trials?

SOLUTION: The probability of rolling a six in one trial is

$$p = \frac{1}{6}$$

and the number of rolls is

$$k = 30$$

therefore

$$E_n = pk$$

$$= \frac{1}{6}\,(30)$$

$$= 5$$

In words, the player would expect to roll a six five times in thirty rolls.

EXAMPLE: If a box contained seven numbered slips of paper, each numbered differently, how many times would a man expect to draw a single selected number slip, if he returned the drawn slip after each draw and he made a total of seventy draws?

SOLUTION: The probability of drawing the selected number slip in one drawing is

$$p = \frac{1}{7}$$

and the number of draws is

$$k = 70$$

therefore

$$E_n = pk$$

$$= (\frac{1}{7})\,70$$

$$= 10$$

Note: When the product of pk is not an integer, we will use the nearest integer to pk.

Value Expectations

We will define value expectation as follows: If, in the event of a successful result, a person is to receive m value and p is the probability of success of that event, then mp is his value expectation.

If you attended a house party where a door prize of $5.00 was given and ten people attended the party, what would be your expectation? In this case, instead of using k for expected number we use m for expected value. That is

$$E_V = pm$$

where

$$p = \text{probability of success}$$

$$m = \text{value of prize}$$

and

$$E_V = \text{expected value}$$

Then, by substitution

$$E_V = pm$$

$$= (\frac{1}{10}) \$5.00$$

$$= \$.50$$

EXAMPLE: In a game, a wheel is spun and when the wheel stops a pointer indicates one of the digits 1, 2, 3, 4, 5, 6, 7, or 8. The prize for winning is $16.00. If a person needed a 6 to win, calculate the following:
(a) What is his probability of winning?
(b) What is his value expectation?

SOLUTION

(a) $p = \dfrac{1}{8}$

(b) $p = \dfrac{1}{8}$ and m = $16.00

therefore

$$E_v = pm = (\dfrac{1}{8})\ \$16.00$$

$$= \$2.00$$

Practice Problems

1. When a store opened, each person who made a purchase was given one ticket on a chance for a door prize of $400. At the close of the day 2,000 people had registered.
    (a) If you made one purchase, what is your expectation?
    (b) If you made 5 purchases, what is your expectation?

2. Each person at a Bingo game purchased a fifty-cent chance for the jackpot of twenty dollars. If fifty people purchased chances, what is each person's
    (a) probability of winning?
    (b) probability of not winning?
    (c) expectation?

Answers

1. (a) $.20
   (b) $1.00

2. (a) $\frac{1}{50}$

   (b) $\frac{49}{50}$

   (c) $.40

Compound Probabilities

The probabilities to this point have been single events. In the discussion on compound probabilities, events which may affect others will be covered. The word "may" is used because included with dependent events and mutually exclusive events is the independent event.

Independent Events

Two events are said to be independent if the occurrence of one has no effect on the occurrence of the other.

When two coins are tossed at the same time or one after the other, whether one falls heads or tails has no effect on the way the second coin falls. Suppose we call the coins A and B. There are four ways in which the coins may fall, as follows:

1. A and B fall heads.
2. A and B fall tails.
3. A falls heads and B falls tails.
4. A falls tails and B falls heads.

The probability of any one way for the coins to fall is calculated as follows:

$$s = 1$$

and

$$n = 4$$

therefore

$$p = \frac{1}{4}$$

This probability may be determined by considering the product of the separate probabilities; that is

p that A falls heads is $\frac{1}{2}$

p that B falls heads is $\frac{1}{2}$

and the probability that both fall heads is

$$\frac{1}{2} \cdot \frac{1}{2} = \frac{1}{4}$$

In other words, when two events are independent, the probability that one and then the other will occur is the product of their separate probabilities.

EXAMPLE: A box contains three red marbles and seven green marbles. If a marble is drawn, then replaced and another marble is drawn, what is the probability that both marbles are red?

SOLUTION: Two solutions are offered. First, there are, by the principle of choice, $10 \cdot 10$ ways in which two marbles can be selected. There are three ways the red marble may be selected on the first draw and three ways on the second draw and by the principle of choice there are $3 \cdot 3$ ways in which a red marble may be drawn on both trials. Then the required probability is

$$p = \frac{9}{100}$$

The second solution, using the product of independent events, follows: The probability of drawing a red ball on the first draw is $\frac{3}{10}$ and the probability of drawing a red ball on the second draw is $\frac{3}{10}$. Therefore, the probability of drawing a red ball on both draws is the product of the separate probabilities

$$p = \frac{3}{10} \cdot \frac{3}{10} = \frac{9}{100}$$

Practice Problems

1. If a die is tossed twice, what is the probability of a two up followed by a three up?

2. A box contains two white, three red, and four blue marbles. If after each selection the marble is replaced, what is the probability of drawing, in order:
    (a) a white then a blue marble?
    (b) a blue then a red marble?
    (c) a white, a red, then a blue marble?

Answers

1. $\frac{1}{36}$

2. (a) $\frac{8}{81}$

(b) $\dfrac{4}{27}$

(c) $\dfrac{8}{243}$

Dependent Events

In some cases one event is dependent on another. That is, two or more events are said to be dependent if the occurrence or nonoccurrence of one of the events affects the probabilities of occurrence of any of the others.

Consider that two or more events are dependent. If p_1 is the probability of a first event, p_2 the probability that after the first happens the second will occur, p_3 the probability that after the first and second have happened the third will occur, etc., then the probability that all events will happen in the given order is the product $p_1 \cdot p_2 \cdot p_3 \cdots$

EXAMPLE: A box contains three white marbles and two black marbles. What is the probability that in two draws both marbles drawn will be black. The first marble drawn is not replaced.

SOLUTION: On the first draw the probability of drawing a black marble is

$$p_1 = \frac{2}{5}$$

and on the second draw the probability of drawing a black marble is

$$p_2 = \frac{1}{4}$$

97

then the probability of drawing both black marbles is

$$p = p_1 \cdot p_2$$

$$= \frac{2}{5} \cdot \frac{1}{4}$$

$$= \frac{1}{10}$$

EXAMPLE: Slips numbered one through nine are placed in a box. If two slips are drawn, what is the probability that
(a) both are odd?
(b) both are even?

SOLUTION:
(a) The probability that the first is odd is

$$p_1 = \frac{5}{9}$$

and the probability that the second is odd is

$$p_2 = \frac{4}{8}$$

therefore, the probability that both are odd is

$$p = p_1 \cdot p_2$$

$$= \frac{5}{9} \cdot \frac{4}{8}$$

$$= \frac{5}{18}$$

(b) The probability that the first is even is

$$p_1 = \frac{4}{9}$$

and the probability that the second is even is

$$p_2 = \frac{3}{8}$$

therefore, the probability that both are even is

$$p = p_1 \cdot p_2$$

$$= \frac{4}{9} \cdot \frac{3}{8}$$

$$= \frac{1}{6}$$

A second method of solution involves the use of combinations.

(a) There are a total of nine slips taken two at a time and there are five odd slips taken two at a time, therefore

$$p = \frac{{}_5C_2}{{}_9C_2}$$

$$= \frac{\dfrac{5!}{2! \ (3!)}}{\dfrac{9!}{2! \ (7!)}}$$

$$= \frac{5}{18}$$

(b) There are a total of ${}_9C_2$ choices and four even slips taken two at a time, therefore

$$p = \frac{{}_4C_2}{{}_9C_2}$$

$$= \frac{1}{6}$$

Practice Problems

In the following problems assume that no replacement is made after each selection.

1. A box contains five white and six red marbles. What is the probability of successfully drawing, in order, a red marble then a white marble?

2. A bag contains three red, two white, and six blue marbles. What is the probability of drawing, in order, two red, one blue and two white marbles?

3. There are fifteen airmen in the line crew. They must take care of the coffee mess and line shack cleanup. They put slips numbered 1 through 15 in a hat and decide that any who draws a number divisible by 5 will be assigned the coffee mess and any who draws a number divisible by 4 will be assigned cleanup. The first person draws a 4, the second a 3, and the third an 11. As fourth person to draw, what is the probability that you will:

(a) be assigned the coffee mess?

(b) be assigned the cleanup?

Answers

1. $\frac{3}{11}$

2. $\frac{1}{770}$

3. (a) $\frac{1}{4}$

 (b) $\frac{1}{6}$

Mutually Exclusive Events

Two or more events are called mutually exclusive if the occurrence of any one of them excludes the occurrence of the others. The probability of occurrence of some one of two or more mutually exclusive events is the sum of the probabilities of the individual events.

It sometimes happens that when one event has occurred, the probability of another event is excluded, it being understood that we are referring to the same given occasion or trial.

For example, throwing a die once can yield a 5 or 6, but not both, in the same toss. The probability that either a 5 or a 6 occurs is the sum of their individual probabilities.

$$p = p_1 + p_2$$
$$= \frac{1}{6} + \frac{1}{6}$$
$$= \frac{1}{3}$$

EXAMPLE: From a bag containing five white balls, two black balls, and eleven red balls, one ball is drawn. What is the probability that it is either black or red?

SOLUTION: There are eighteen ways in which the draw can be made. There are two black ball choices and eleven red ball choices which are favorable, or a total of thirteen favorable choices. Then, the probability of success is

$$p = \frac{13}{18}$$

Since drawing a red ball excludes the drawing of a black ball, and vice versa, the two events are mutually exclusive. Then, the probability of drawing a black ball is

$$p_1 = \frac{2}{18}$$

and the probability of drawing a red ball is

$$p_2 = \frac{11}{18}$$

Therefore the probability of success is

$$p = p_1 + p_2$$

$$= \frac{2}{18} + \frac{11}{18} = \frac{13}{18}$$

EXAMPLE: What is the probability of drawing either a king, a queen, or a jack from a deck of playing cards?

SOLUTION: The individual probabilities are

$$\text{king} = \frac{4}{52}$$

$$\text{queen} = \frac{4}{52}$$

$$\text{jack} = \frac{4}{52}$$

Therefore the probability of success is

$$p = \frac{4}{52} + \frac{4}{52} + \frac{4}{52}$$

$$= \frac{12}{52}$$

$$= \frac{3}{13}$$

EXAMPLE: What is the probability of rolling a die twice and having a 5 and then a 3 show or having a 2 and then a 4 show?

SOLUTION: The probability of having a 5 and then a 3 show is

$$p_1 = \frac{1}{6} \cdot \frac{1}{6}$$

$$= \frac{1}{36}$$

and the probability of having a 2 and then a 4 show is

$$p_2 = \left(\frac{1}{6}\right)\left(\frac{1}{6}\right)$$

$$= \frac{1}{36}$$

Then, the probability of either p_1 or p_2 is

$$p = p_1 + p_2$$

$$= \frac{1}{36} + \frac{1}{36}$$

$$= \frac{1}{18}$$

Practice Problems

1. When tossing a coin, what is the probability of getting either a head or a tail?

2. A bag contains twelve blue, three red, and four white marbles. What is the probability of drawing:

    (a) in one draw, either a red or a white marble?

    (b) in one draw, either a red, white, or blue marble?

    (c) in two draws, either a red marble followed by a blue marble or a red marble followed by a red marble?

3. What is the probability of getting a total of at least 10 points in rolling two dice?

(HINT: You want either a total of 10, 11, or 12.)

Answers

1. 1

2. (a) $\frac{7}{19}$

   (b) 1

   (c) $\frac{7}{57}$

3. $\frac{1}{6}$

Empirical Probabilities

Among the most important applications of probability are those in situations where we cannot list all possible outcomes. To this point we have considered problems in which the probabilities could be obtained from situations in terms of equally likely results.

Because some problems are so complicated for analysis we can only estimate probabilities from experience and observation. This is empirical probability.

In modern industry probability now plays an important role in many activities. Quality control and reliability of a manufactured article have become extremely important considerations in which probability is used.

Relative Frequency of Success

We define relative frequency of success as follows. After N trials of an event have been made, of which S trials are success, then the relative frequency of success is

$$P = \frac{S}{N}$$

Experience has shown that empirical probabilities, if carefully determined on the basis of adequate statistical samples, can be applied to large groups with the result that probability and relative frequency are approximately equal. By adequate samples we mean a large enough sample so that accidental runs of "luck," both good and bad, cancel each other. With enough trials, predicted results and actual results agree quite closely. On the other hand, applying a probability ratio to a single individual event is virtually meaningless.

For example, table 17-1 shows a small number of weather forecasts from April 1 to April 10. The actual weather on the dates is also given.

Observe that the forecasts on April 1, 3, 4, 6, 7, 8, and 10 were correct. We have observed ten outcomes. The event of a correct forecast has occurred seven times. Based on this information we might say that the probability for future forecasts being true is $\frac{7}{10}$. This number is the best estimate that we can make from the given information. In this case, since we have observed such a small number of outcomes, it would not be correct to say that the estimate of P is dependable. A great many more cases should be used if we expect to make a good estimate of the probability that a weather forecast will be accurate. There are a great many factors which affect the accuracy of a weather forecast. This example merely indicates something about how successful a particular weather office has been in making weather forecasts.

Table 17-1. —Weather forecast.

Date	Forecast	Actual weather	Did the actual forecasted event occur?
1	Rain	Rain	yes
2	Light showers	Sunny	No
3	Cloudy	Cloudy	Yes
4	Clear	Clear	Yes
5	Scattered showers	Warm and sunny	No
6	Scattered showers	Scattered showers	Yes
7	Windy and cloudy	Overcast and windy	Yes
8	Thundershowers	Thundershowers	Yes
9	Clear	Cloudy and rain	No
10	Clear	Clear	Yes

Another example may be drawn from industry. Many thousands of articles of a certain type are manufactured. The company selects 100 of these articles at random and subjects them to very careful tests. In these tests it is found that 98 of the articles meet all measurement requirements and perform satisfactorily. This suggests that $\frac{98}{100}$ is a measure of the reliability of the article.

One might expect that about 98% of all of the articles manufactured by this process will be satisfactory. The probability (measure of chance) that one of these articles will be satisfactory might be said to be 0.98.

This second example of empirical probability is different from the first example in one very important respect. In the first example all of the possibilities could be listed and in the second example we could not do so. The selection of a sample and its size is a problem of statistics.

Considered from another point of view, statistical probability can be regarded as relative frequency.

EXAMPLE: In a dart game, a player hit the bull's eye 3 times out of 25 trials. What is the statistical probability that he will hit the bull's eye on the next throw?

SOLUTION:

$$N = 25$$

and

$$S = 3$$

hence

$$P = \frac{3}{25}$$

EXAMPLE: Using table 17-2, what is the probability that a person 20 years old will live to be 50 years old?

Table 17-2. —Mortality table (based on 100,000 indi‑
viduals 1 year of age).

Age	Number of people
5	98,382
10	97,180
15	96,227
20	95,148
25	93,920
30	92,461
35	90,655
40	88,334
45	85,255
50	81,090
55	75,419
60	67,777
65	57,778
70	45,455
75	31,598
80	18,177
85	7,822
90	2,158

SOLUTION: Of 95,148 persons at age 20, 81,090 survived to age 50. Hence

$$P = \frac{81,090}{95,148}$$

$$= 0.852$$

EXAMPLE: How many times would a die be expected to land with 5 or 6 showing in 20 trials?

SOLUTION: The probability of a 5 or 6 showing is

$$p = \frac{1}{3}$$

The relative frequency is approximately equal to the probability

$$P \approx p$$

therefore

$$P = \frac{S}{N}$$

where

$$P = \frac{1}{3}$$

$$S = ?$$

and

$$N = 20$$

Rearranging and substituting, find that

$$P = \frac{S}{N}$$

$$\frac{1}{3} = \frac{S}{20}$$

$$S = \frac{20}{3}$$

$$= 7$$

(NOTE: The number observed in an experiment may differ from that predicted; therefore, the results may be taken to the nearest integer.)

Practice Problems

1. In a construction crew there are six electricians and 38 other workers. How many electricians would you expect to choose if you chose one man each day of a week for your helper?

2. How many times would a tossed die be expected to turn up 3 or less in thirty tosses?

3. Using table 17-2, find the probability that a person whose age is 30 will live to age 60.

Answers

1. 1
2. 15
3. 0.733

Data Processing

Data processing is an extremely large and complex field and applications are usually made for individual situations. For a general understanding of how data processing can be related to probability and statistics, the functional operation of the computer is needed.

Computer Operations

High speed computers are used in data processing because they are able to solve problems in seconds where humans may require months and even years to solve the same problem.

A problem which arises, when using a computer, is how the human can communicate with the computer. This communication is a function of mathematics. We refer to this form of mathematics as computer-oriented mathematics.

Digital computers are high speed adding machines. To perform multiplication they make repeated additions. To perform subtraction the addition sequence may be reversed, and division is the process of repeated subtraction.

An example of an operation of a digital computer is finding the square root of 36. We know that the sum of the first n odd integers is equal to n^2. The computer is programed to subtract successive odd integers from 36 until zero is reached. That is,

$$36 - 1 - 3 - 5 - 7 - 9 - 11 = 0$$

When zero is reached, the computer then counts the number of odd integers it has subtracted and this sum is the square root of 36.

Application to Probability

Many problems may be solved on computers with the use of the proper mathematical model. The mathematical model may include any of the variables and the probability with which they occur. The collection of statistical data plays an important part in building a mathematical model for the computer.

The mathematical model may be used to determine probabilities by the use of computers. That is, the computer will "play a game" many times and give a result comparable to many trials for determining probabilities.

To understand how a game of chance may be used to produce a useful result, consider the problem of determining the product of 3/8 and 2/3. Place eight ping-pong balls of which three are coated with a conducting material in one container. Place three other ping-pong balls of which two are coated with the conducting material in a second container. A trial consists of a detecting head from the computer touching a ball in each container. If both balls are coated a point is registered; if not, a zero is registered. The number of points registered divided by the total trials, if the number of trials is large, will closely approximate the fraction 6/24.

The reasoning for the result is that the probability of touching a coated ball in the first container is 3/8 and the probability of touching a coated ball in the second container

113

is 2/3. The events of touching a coated ball in the first container and one in the second container are independent, and the probability of both balls being coated is the product of the individual probabilities. This is exactly what we set out to determine.

The preceding example is extremely simplified, in comparison with the complexity of the actual statistical problems solved by computers. However, it does serve to indicate some of the possibilities of computer-oriented mathematics.

Usage of Statistical Data

Suppose a squadron, through years of operation, has accumulated statistical data on the operation of an aircraft. By using a computer, the probability of the failure of an engine can be determined from the many bits of information regarding individual parts of the engine. A related problem is to determine how the engine failure probability can be decreased.

While changing all of the components of an aircraft engine to try to improve efficiency is unsound, the mathematical model, from a selected group of experiments, may be used to predict the change in efficiency for any combination of component changes.

One solution is by trial and error, but this would take years of time. Another solution is to make a few flights using various configurations of the engine and then use the computer to simulate years of operation. Researchers would then determine the probability of failure, based on the new statistical data obtained from the few flights. By this high speed determination of the failure of the engine with different combinations of new parts, the optimum design of the modified engine may be determined.

Additional Solved Problems

In a survey carried out in a school snack shop, the following results were obtained. Of 100 boys questioned, 78 liked sweets, 74 liked ice cream, 53 liked cake, 57 liked both sweets and ice cream, 46 liked both sweets and cake, and only 31 boys liked all three. If all the boys interviewed liked at least one item, draw a Venn diagram to illustrate the results. How many boys like both ice cream and cake?

SOLUTION:

A Venn diagram is a pictorial representation of the relationship between sets. A set is a collection of objects. The number of objects in a particular set is the cardinality of a set.

To draw a Venn diagram, we start with the following picture:

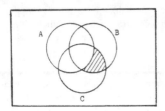

Each circle represents set A, B, or C, respectively. Let

A = set of boys who like ice cream,

B = set of boys who like cake,

C = set of boys who like sweets.

The sections of overlap between circles represents the members of one set who are also members of another set. For example, the shaded region in the picture indicates the set of boys who are in sets B and C but not A. This is the set of boys who like both cake and sweets but not ice cream. The inner section common to all three circles indicates the set of boys who belong to all three sets simultaneously.

We wish to find the number of boys who liked both ice cream and cake. Let us label the sections of the diagram with the cardinality of these sections. The cardinality of the region common to all three sets is the number of boys who liked all three items, or 31.

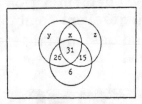

The number of boys who like ice cream and sweets was 57. Of these 57, 31 like all three, leaving 26 boys in set A and set C but not set B. Similarly, there are 15 boys in B and C, but not in A. There are $78 - 26 - 31 - 15 = 6$ boys in C but not in A or B.

Let x = number of boys who are in A and B but not C

y = number of boys who are in A but not B or C

z = number of boys who are in B but not A or C

We know that the sum of all the labeled areas is 100 or

$$26 + 31 + 15 + 6 + x + y + z = 100$$
$$78 + x + y + z = 100$$

Also, there are 74 boys total in set A, or

$$x + y + 31 + 26 = 74 ,$$

and 53 total in set B, or

$$x + z + 46 = 53.$$

Combining:
$$x + y + z = 100 - 78 = 22$$
$$x + y = 74 - 57 = 17$$
$$x + z = 53 - 46 = 7$$

Subtracting the second equation from the first gives $z = 5$, implying $x = 2$ and $y = 15$. Our answer is the number of boys in sets A and $B = x + 31 = 33$.

Of 37 men and 33 women, 36 are teetotalers. Nine of the women are non-smokers and 18 of the men smoke but do not drink. Seven of the women and 13 of the men drink but do not smoke. How many, at most, both drink and smoke?

SOLUTION:

A = set of all smokers

B = set of all drinkers

C = set of all women

D = set of all men

We construct two Venn diagrams and label them in the following way:

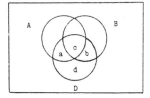 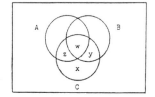

Each section on the graph indicates a subset of the group of men and women. For example, the section labeled "z" is the subset including all women who smoke but do not drink. The section labeled "b" is the subset including all men who drink but do not smoke.

In addition to labels, these letters also will indicate the cardinality — the number of objects — in the subset. We are told there are 37 men; thus, $a + b + c + d = 37$. There are 33 women; thus, $x + y + z + w = 33$. There are 9 women non-smokers which includes $x + y$. The number of non-drinking, smoking men is $d = 18$.

Similarly, $x + z + a + d = 36$, the non-drinkers
$b = 13$, the drinking, non-smoking men
$y = 7$, the drinking, non-smoking women.

Collecting all these equations, we wish to find the maximum value of $c + w$, the number of drinkers and smokers.

$$x + z + a + d = 36, \quad a + b + c + d = 37$$
$$b = 13, \, d = 18, \quad x + y + z + w = 33$$
$$y = 7, \quad x + y = 9 \ .$$

Substituting we see that

$$x + y = x + 7 = 9 \text{ or } x = 2.$$

From this we have

$$2 + z + a + 18 = 36 \quad a + 13 + c + 18 = 37$$
$$a + z = 16 \quad a + c = 6$$
$$2 + 7 + z + w = 33$$
$$z + w = 24$$

We now solve for $c + w$.

$$a = 6 - c \text{ and thus } z + 6 - c = 16 \text{ or } z - c = 10$$
$$c = z - 10 \text{ and } w = 24 - z$$
thus, $c + w = z - 10 + 24 - z = 14.$

The maximum number of drinkers and smokers is 14.

What is the probability of throwing a six with a single die?

SOLUTION:

The die may land in any of six ways:
1, 2, 3, 4, 5, or 6.
The probability of throwing a six is

$$P\{6\} = \frac{\text{number of ways to get a six}}{\text{number of ways the die may land}}$$

Thus ,
$$P\{6\} = \frac{1}{6}.$$

A deck of playing cards is thoroughly shuffled and a card is drawn at random from the deck. What is the probability that the card drawn is the ace of diamonds?

SOLUTION:

The probability of this event occurring is

$$\frac{\text{number of ways this event can occur}}{\text{number of possible outcomes}}.$$

In our case there is only one way this event can occur, for there is only one ace of diamonds and there are 52 possible outcomes (for there are 52 cards in the deck). Hence, the probability that the card drawn is the ace of

diamonds is $\frac{1}{52}$.

A box contains seven red, five white, and four black balls. One ball is drawn at random. What is the probability of drawing a red ball? A black ball?

SOLUTION:

There are $7 + 5 + 4 = 16$ balls in the box. The probability of drawing a red ball is

$$P\{R\} = \frac{\text{number of possible ways of drawing a red ball}}{\text{the number of ways of drawing any ball}}$$

$$P\{R\} = \frac{7}{16}.$$

Similarly, the probability of drawing a black ball is

$$P\{B\} = \frac{\text{number of possible ways of drawing a black ball}}{\text{number of ways of drawing any ball}}$$

Thus, $P\{B\} = \frac{4}{16} = \frac{1}{4}.$

Find the probability of drawing a black card in a single random draw from a well-shuffled deck of ordinary playing cards.

SOLUTION:

There are 52 cards and since the cards are well-shuffled, each card is assumed to be equally likely to be drawn. There are 26 black cards in the deck, and thus the number of outcomes leading to a black card being drawn is 26. Therefore,

$$P \text{ \{drawing a black card\}} = \frac{26}{52} = \frac{1}{2}.$$

Find the probability of drawing a spade on a single random draw from a well-shuffled deck of cards.

SOLUTION:

There are 52 possible outcomes to the experiment of drawing a card. There are 13 spades in a deck and hence 13 possible outcomes to the experiment which lead to drawing a spade.

Thus, $P \text{ \{drawing a spade\}} = \frac{13}{52} = \frac{1}{4}.$

What is the probability of obtaining a sum of seven in a single throw of a pair of dice?

SOLUTION:

There are $6 \times 6 = 36$ outcomes which could result from two dice being thrown, as shown in the accompanying figure.

1,1	1,2	1,3	1,4	1,5	1,6
2,1	2,2	2,3	2,4	2,5	2,6
3,1	3,2	3,3	3,4	3,5	3,6
4,1	4,2	4,3	4,4	4,5	4,6
5,1	5,2	5,3	5,4	5,5	5,6
6,1	6,2	6,3	6,4	6,5	6,6

The number of possible ways that a seven will appear are circled in the figure. Let us call this set B. Thus, $B = \{ (1, 6), (2, 5), (3, 4), (4, 3), (5, 2), (6, 1) \}$.

There are six elements in B, so $P\{7\} = P\{B\} = \dfrac{6}{36} = \dfrac{1}{6}$.

In a single throw of a single die, find the probability of obtaining either a two or a five.

SOLUTION:

In a single throw, the die may land in any of six ways:
$$\{1, 2, 3, 4, 5, 6\} .$$
The probability of obtaining a two is

$$P\{2\} = \frac{\text{number of ways of obtaining a two}}{\text{number of ways the die may land}} , \; P\{2\} = \frac{1}{6} .$$

Similarly, the probability of obtaining a five is

$$P\{5\} = \frac{\text{number of ways of obtaining a five}}{\text{number of ways the die may land}} , \; P\{5\} = \frac{1}{6} .$$

As it is impossible for the single throw to result in a two and a five simultaneously, the two events are mutually exclusive. The probability that either one of two mutually exclusive events will occur is the sum of the probabilities of the separate events. Thus, the probability of obtaining either a two or a five, $P\{2 \text{ or } 5\}$, is

$$P\{2\} + P\{5\} = \frac{1}{6} + \frac{1}{6} = \frac{2}{6} = \frac{1}{3} .$$

If a card is drawn from a deck of playing cards, what is the probability that it will be a jack or a ten?

SOLUTION:

The probability of the union of two mutually exclusive events is $P\{A \cup B\} = P\{A\} + P\{B\}$. Here the symbol "$\cup$" stands for "or."

In this particular example, we only select one card at a time. Thus, we either choose a jack "or" a ten. Since no single selection can simultaneously be a jack and a ten, the two events are mutually exclusive.

$P\{\text{jack or ten}\} = P\{\text{jack}\} + P\{\text{ten}\}$.

$$P\{\text{jack}\} = \frac{\text{number of ways to select a jack}}{\text{number of ways to choose a card}} = \frac{4}{52} = \frac{1}{13} \ .$$

$$P\{\text{ten}\} = \frac{\text{number of ways to select a ten}}{\text{number of ways to choose a card}} = \frac{4}{52} = \frac{1}{13} \ .$$

$$P\{\text{jack or a ten}\} = P\{\text{jack}\} + P\{\text{ten}\} = \frac{1}{13} + \frac{1}{13} = \frac{2}{13} \ .$$

Suppose that we have a bag containing two red balls, three white balls, and six blue balls. What is the probability of obtaining a red or a white ball on one draw?

SOLUTION:

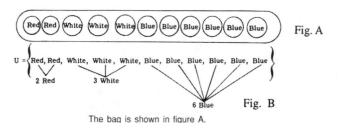

The bag is shown in figure A.

The probability of drawing a red ball is $\frac{2}{11}$, and the probability of drawing a white ball is $\frac{3}{11}$. Since drawing a red and drawing a white ball are mutually exclusive events, the probability of drawing a red or a white ball is the probability of drawing a red ball plus the probability of drawing a white ball, so

$$P \text{ {red or white}} = \frac{2}{11} + \frac{3}{11} = \frac{5}{11} .$$

Note: In the above example the probability of drawing a blue ball would be $\frac{6}{11}$. Therefore, the sum of the probability of a red ball, the probability of a white ball, and the probability of a blue ball is $\frac{2}{11} + \frac{3}{11} + \frac{6}{11} = 1$.

If there are no possible results that are considered favorable, then the probability P {F} is obviously 0. If every result is considered favorable, then P {F} = 1. Hence, the probability P {F} of a favorable result F always satisfies the inequality

$$0 \leq P \text{{F}} \leq 1 .$$

A bag contains four white balls, six black balls, three red balls, and eight green balls. If one ball is drawn from the bag, find the probability that it will be either white or green.

SOLUTION:

The probability that it will be either white or green is

$$P \text{ {white or green}} = P \text{ {white}} + P \text{ {green}} .$$

This is true because if we are given two mutually exclusive events A and B, then P {A or B} = P {A} + P {B}. Note that two events, A and B, are mutually exclusive events if their intersection is the null or empty set. In this case the intersection of choosing a white ball and of choosing a green ball is the empty set. There are no elements in common. Now

$$P \text{ \{white\}} = \frac{\text{number of ways to choose a white ball}}{\text{number of ways to select a ball}}$$

$$= \frac{4}{21} \text{ , and}$$

$$P \text{ \{green\}} = \frac{\text{number of ways to choose a green ball}}{\text{number of ways to select a ball}}$$

$$= \frac{8}{21} \text{ .}$$

Thus,

$$P \text{ \{white or green\}} = \frac{4}{21} + \frac{8}{21} = \frac{12}{21} = \frac{4}{7} \text{ .}$$

Find the probability of obtaining a sum of either six or seven in a single toss of two dice.

SOLUTION:

Let A = the event that a sum of six is obtained in a toss of two dice

B = the event that a sum of seven is obtained in a toss of two dice.

Then the probability of obtaining either a six or a seven in a single toss of two dice is

$$P \text{ \{A or B\}} = P \text{ \{A } \cup \text{ B\}} \text{ .}$$

The union symbol "$\cup$" means that A and/or B can occur. Now $P \text{ \{A } \cup \text{ B\}}$ = $P \text{ \{A\}} + \text{ \{B\}}$ if A and B are mutually exclusive. Two or more events are said to be mutually exclusive if the occurrence of any one of them excludes the occurrence of the others. In this case, we cannot obtain a six and a seven in a single toss of two dice. Thus, A and B are mutually exclusive.

Note: There are 36 different tosses of two dice.

A = a six is obtained in a toss of two dice
$= \{ (1, 5), (2, 4), (3, 3), (4, 2), (5, 1) \}$
B = a seven is obtained in a toss of two dice
$= \{ (1, 6), (2, 5), (3, 4), (4, 3), (5, 2), (6, 1) \}.$

$$P\{A\} = \frac{\text{number of ways to obtain a six in a toss of two dice}}{\text{number of ways to toss two dice}}$$

$$= \frac{5}{36}$$

$$P\{B\} = \frac{\text{number of ways to obtain a seven in a toss of two dice}}{\text{number of ways to toss two dice}}$$

$$= \frac{6}{36} = \frac{1}{6}$$

Therefore, $P\{A \cup B\} = P\{A\} + \{B\} = \frac{5}{36} + \frac{6}{36} = \frac{11}{36}$.

A coin is tossed nine times. What is the total number of possible outcomes of the nine–toss experiment? How many elements are in the subset "six heads and three tails"? What is the probability of getting exactly six heads and three tails in nine tosses of this unbiased coin?

SOLUTION:

There are two possible outcomes for each toss, and thus $\underbrace{2 \times 2 \times \ldots \times}_{\text{nine terms}} = 2^9$ possible outcomes in nine tosses, or 512 outcomes.

To count the number of elements in the subset "six heads and three tails" is equivalent to counting the number of ways six objects can be selected from nine. These objects will then be labeled "heads" and the remaining three objects will be labeled "tails." There are

$$\binom{9}{6} = \frac{9!}{6! \; 3!} = 84$$

ways to do this and hence the probability of observing this configuration is

$$\frac{\text{number of ways six heads and three tails can occur}}{\text{total possible outcomes}}$$

$$= \frac{\binom{9}{6}}{2^9} = \frac{84}{512} = 0.164$$

Suppose a die has been loaded so that the $\boxed{\bullet}$ face lands uppermost three times as often as any other face while all the other faces occur equally often. What is the probability of a $\boxed{\bullet\bullet}$ on a single toss? What is the probability of a $\boxed{\bullet}$?

SOLUTION:

Let p equal the probability of the $\boxed{\bullet}$ face landing uppermost. We know that $P\{\boxed{\bullet}\} = 3\,P$ {any other face}. We also know that faces with j dots, $j = 2, 3, 4, 5, 6$ occur equally often. Thus,

$$\sum_{j=1}^{6} P\,\{j \text{ dots}\} = 1,$$

and $\quad P$ {one dot} +

$$\sum_{j=2}^{6} P\,\{j \text{ dots}\} = 1$$

or $\quad p + 5\left(\frac{1}{3}\,P\right) = 1$.

Thus, $p = \frac{3}{8}$ and P {two dots} $= \frac{1}{3}\,P = \frac{1}{8}$.

The probability of a $\boxed{\bullet}$ is $\frac{3}{8}$.

In a single throw of a pair of dice, find the probability of obtaining a total sum of four or less.

SOLUTION:

Each die may land in six ways. By the Fundamental Principle of Counting the pair of dice may thus land in $6 \times 6 = 36$ ways:

1,1	1,2	1,3	1,4	1,5	1,6
2,1	2,2	2,3	2,4	2,5	2,6
3,1	3,2	3,3	3,4	3,5	3,6
4,1	4,2	4,3	4,4	4,5	4,6
5,1	5,2	5,3	5,4	5,5	5,6
6,1	6,2	6,3	6,4	6,5	6,6

Let us call the possible outcomes which are circled above set A. Then the elements of set A, $A = \{ (1, 1), (1, 2), (1, 3), (2, 1), (2, 2), (3, 1) \}$, are all the possible ways of obtaining a total sum of four or less.

The probability of obtaining four or less is

$$P \{(x + y) \le 4\} = \frac{\text{no. of ways of obtaining four or less } \left(\text{no. of elements in set } A\right)}{\text{no. of ways the dice may land}}$$

$$= \frac{6}{36} = \frac{1}{6}$$

A card is drawn at random from a deck of cards. Find the probability that at least one of the following three events will occur:

Event A: a heart is drawn.
Event B: a card which is not a face card is drawn.
Event C: the number of spots (if any) on the drawn card is evenly divisible by three.

SOLUTION:

Let $A \cup B \cup C =$ the event that at least one of the three events above will occur. We wish to find P $\{A \cup B \cup C\}$, the probability of the event $A \cup B \cup C$. Let us count the number of ways that at least A, B, or C will occur. There are 13 hearts, 40 non-face cards, and 12 cards such that the number of spots is divisible by three (cards numbered 3, 6, or 9 are all divisible by three and there are four suits each with three such cards, $3 \times 4 = 12$). If we add $40 + 13 + 12$ we will have counted too many times. There are 10 cards which are hearts and non-face cards, three cards divisible by three and hearts, and 12 cards which are non-face cards and divisible by three. We must subtract each of these from our total of $40 + 13 + 12$ giving $40 + 13 + 12 - 10 - 3 - 12$. But we have subtracted too much; we have subtracted the three cards which are hearts and non-face cards and divisible by three. We must add these cards to our total, making

$$P \{A \cup B \cup C\} = \frac{40 + 13 + 12 - 10 - 3 - 12 + 3}{52} = \frac{43}{52} .$$

This counting technique used is called the Principle of Inclusion/Exclusion and is useful for problems of this sort. Notice that

$$P \{A \cup B \cup C\} = \frac{40 + 13 + 12 - 10 - 3 - 12 + 3}{52}$$

$$= \frac{13}{52} + \frac{40}{52} + \frac{12}{52} - \frac{10}{52} - \frac{3}{52} - \frac{12}{52} + \frac{3}{52} .$$

$$= P \{A\} + P \{B\} + P \{C\} - P \{AB\} - P \{AC\} - P \{BC\} + P \{ABC\}, \text{ since}$$

$$P \{A\} = \frac{\text{number of hearts}}{\text{number of cards}} = \frac{13}{52} ,$$

$$P \{B\} = \frac{\text{number of non-face cards}}{\text{number of cards}} = \frac{40}{52} ,$$

$$P \{C\} = \frac{\text{number of cards divisible by three}}{\text{number of cards}} = \frac{12}{52} ,$$

$$P \{AB\} = \frac{\text{number of hearts and non -face cards}}{\text{number of cards}} = \frac{10}{52} ,$$

$$P \{AC\} = \frac{\text{number of hearts and cards divisible by three}}{\text{number of cards}} = \frac{3}{52} ,$$

$$P \{BC\} = \frac{\text{number of number cards divisible by three}}{\text{number of cards}} = \frac{12}{52} , \text{ and}$$

$$P\ \{ABC\} = \frac{3}{52}\ .$$

Find the probability of obtaining on a single throw of a pair of dice a sum total of five, six, or seven.

Define the events A, B, and C as follows:
Event A: a sum total of five is thrown,
Event B: a sum total of six is thrown, and
Event C: a sum total of seven is thrown.

SOLUTION:

Only one of these three events can occur at one time. The occurrence of any one excludes the occurrence of any of the others. Such events are called mutually exclusive. Let $A \cup B \cup C$ = the event that a sum total of five, six, or seven is observed. Then $P\ \{A \cup B \cup C\} = P\ \{A\} + P\ \{B\} + P\ \{C\}$, because the events are mutually exclusive. Referring to a previous table, we see that

$P\ \{A\} = \frac{4}{36}$, $P\ \{B\} = \frac{5}{36}$, and $P\ \{C\} = \frac{6}{36}$. Therefore,

$$P\ \{A \cup B \cup C\} = \frac{4}{36} + \frac{5}{36} + \frac{6}{36} = \frac{15}{36} = \frac{5}{12}\ .$$

What is the probability of obtaining a five on each of two successive rolls of a balanced die?

SOLUTION:

We are dealing with separate rolls of a balanced die. The two rolls are independent, and therefore we invoke the following multiplication rule: the probability of obtaining any particular combination in two or more independent trials will be the product of their individual probabilities. The probability of obtaining a five on any single toss is $\frac{1}{6}$, and by the multiplication rule

$$P \{5 \text{ and } 5\} = \frac{1}{6} \times \frac{1}{6} = \frac{1}{36}.$$

If a pair of dice is tossed twice, find the probability of obtaining a sum total of five on both tosses.

SOLUTION:

The ways to obtain five in one toss of the two dice are
(1, 4), (4, 1), (3, 2), and (2, 3) .

Hence, we can throw five in one toss in four ways. Each die has six faces and there are six ways for a die to fall. Then the pair of dice can fall in $6 \times 6 = 36$ ways. The probability of throwing five in one toss is

$$\frac{\text{number of ways to throw a five in one toss}}{\text{number of ways that a pair of dice can fall}} = \frac{4}{36} = \frac{1}{9} .$$

Now the probability of throwing a five on both tosses is
P {throwing five on first toss and throwing five on second toss} .
"And" implies multiplication if events are independent, thus
P {throwing five on first toss and throwing five on second toss}
$= P$ {throwing five on first toss} $\times P$ {throwing five on second toss},
since the results of the two tosses are independent. Consequently, the probability of obtaining five on both tosses is

$$\left(\frac{1}{9}\right) \times \left(\frac{1}{9}\right) = \frac{1}{81} .$$

A penny is to be tossed three times. What is the probability that there will be two heads and one tail?

SOLUTION:

We start this problem by constructing a set of all possible outcomes:

We can have heads on all three tosses:	(HHH)
heads on the first two tosses, a tail on the third:	(HHT) (1)
a head on the first toss, and tails on the next two:	(HTT)
	(HTH) (2)
	(THH) (3)
	(THT)
	(TTH)
	(TTT)

Hence, there are eight possible outcomes (two possibilities on the first toss × two on the second, and × two on the third = 2 × 2 × 2 = 8).

If the coin is fair, then these outcomes are all equally likely and we assign the probability $\frac{1}{8}$ to each. Now we look for the set of outcomes that produce two heads and one tail. We see there are three such outcomes out of the eight possibilities (numbered (1), (2), (3) in our listing). Hence, the probability of two heads and one tail is $\frac{3}{8}$.

Of the approximately 635,000,000,000 different bridge hands, how many contain a ten–card suit?

SOLUTION:

As before, the number of ways a hand could be dealt can be thought of as the number of ways one can select such a hand. We will select a 13-card bridge hand with ten cards of the same suit in three steps. First, let us select the suit of the 10 cards; there are $\binom{4}{1}$ or 4 ways to do this. Next, let us select 10 of the 13 possible cards in the suit which will be in our bridge hand. There are $\binom{13}{10}$ ways to do this, with

$$\binom{13}{10}=\frac{13!}{10!\times 3!}=\frac{13\times 2\times 11}{3\times 2\times 1}=286 \text{ ways to choose the 10 cards.}$$

Last, we must select the three remaining cards. These three cards must not be in the same suit as the 10 cards we have previously chosen. Instead they must be from the remaining three suits, or 39 cards. There are $\binom{39}{3}$ ways to select these cards and

$$\binom{39}{3}=\frac{39!}{3!\times 36!}=\frac{39\times 38\times 37}{3\times 2\times 1}=9139.$$

By the Fundamental Principle of Counting, the number of different hands containing 10 cards in the same suit is the product of the number of ways in which each of these steps might be carried out. This product is

$$\binom{4}{1}\times\binom{13}{10}\times\binom{39}{3} = 4 \times 286 \times 9{,}139 = 10{,}455{,}016 .\text{This is the}$$

number of all the possible bridge hands which have 10 cards all of the same suit.

We can now calculate the probability of being dealt a hand with 10 cards in the same suit. This probability is the

$$\frac{\text{number of hands with ten cards in the same suit}}{\text{total number of possible bridge hands}} .$$

This equals

$$\frac{\binom{4}{1}\times\binom{13}{10}\times\binom{39}{3}}{\binom{52}{13}}=0.000016 .$$

Available at your local bookstore or order directly from us by sending in coupon below.

MAXnotes®

REA's Literature Study Guides

MAXnotes® are student-friendly. They offer a fresh look at masterpieces of literature, presented in a lively and interesting fashion. **MAXnotes**® offer the essentials of what you should know about the work, including outlines, explanations and discussions of the plot, character lists, analyses, and historical context. **MAXnotes**® are designed to help you think independently about literary works by raising various issues and thought-provoking ideas and questions. Written by literary experts who currently teach the subject, **MAXnotes**® enhance your understanding and enjoyment of the work.

Available **MAXnotes**® include the following:

Absalom, Absalom!
The Aeneid of Virgil
Animal Farm
Antony and Cleopatra
As I Lay Dying
As You Like It
The Autobiography of
    Malcolm X
The Awakening
Beloved
Beowulf
Billy Budd
The Bluest Eye, A Novel
Brave New World
The Canterbury Tales
The Catcher in the Rye
The Color Purple
The Crucible
Death in Venice
Death of a Salesman
The Divine Comedy I: Inferno
Dubliners
The Edible Woman
Emma
Euripides' Medea & Electra
Frankenstein
Gone with the Wind
The Grapes of Wrath
Great Expectations
The Great Gatsby
Gulliver's Travels
Handmaid's Tale
Hamlet
Hard Times
Heart of Darkness

Henry IV, Part I
Henry V
The House on Mango Street
Huckleberry Finn
I Know Why the Caged
    Bird Sings
The Iliad
Invisible Man
Jane Eyre
Jazz
The Joy Luck Club
Jude the Obscure
Julius Caesar
King Lear
Leaves of Grass
Les Misérables
Lord of the Flies
Macbeth
The Merchant of Venice
Metamorphoses of Ovid
Metamorphosis
Middlemarch
A Midsummer Night's Dream
Moby-Dick
Moll Flanders
Mrs. Dalloway
Much Ado About Nothing
Mules and Men
My Antonia
Native Son
1984
The Odyssey
Oedipus Trilogy
Of Mice and Men
On the Road

Othello
Paradise
Paradise Lost
A Passage to India
Plato's Republic
Portrait of a Lady
A Portrait of the Artist
    as a Young Man
Pride and Prejudice
A Raisin in the Sun
Richard II
Romeo and Juliet
The Scarlet Letter
Sir Gawain and the
    Green Knight
Slaughterhouse-Five
Song of Solomon
The Sound and the Fury
The Stranger
Sula
The Sun Also Rises
A Tale of Two Cities
The Taming of the Shrew
Tar Baby
The Tempest
Tess of the D'Urbervilles
Their Eyes Were Watching God
Things Fall Apart
To Kill a Mockingbird
To the Lighthouse
Twelfth Night
Uncle Tom's Cabin
Waiting for Godot
Wuthering Heights
Guide to Literary Terms

RESEARCH & EDUCATION ASSOCIATION
61 Ethel Road W. • Piscataway, New Jersey 08854
Phone: (732) 819-8880   **website: www.rea.com**

Please send me more information about MAXnotes®.

Name _____

Address _____

City _____ State _____ Zip _____

HANDBOOK of
**MATHEMATICAL,
SCIENTIFIC, &
ENGINEERING**
FORMULAS, FUNCTIONS,
GRAPHS, TABLES,
& TRANSFORMS

REA RESEARCH & EDUCATION ASSOCIATION

A particularly useful reference for those in math, science, engineering and other chnical fields. Includes the most-often used formulas, tables, transforms, functions, and aphs which are needed as tools in solving problems. The entire field of special functions also covered. A large amount of scientific data which is often of interest to scientists d engineers has been included.

Available at your local bookstore or order directly from us by sending in coupon below.

REA's **Problem Solvers**

The "PROBLEM SOLVERS" are comprehensive supplemental text-books designed to save time in finding solutions to problems. Each "PROBLEM SOLVER" is the first of its kind ever produced in its field. It is the product of a massive effort to illustrate almost any imaginable problem in exceptional depth, detail, and clarity. Each problem is worked out in detail with a step-by-step solution, and the problems are arranged in order of complexity from elementary to advanced. Each book is fully indexed for locating problems rapidly.

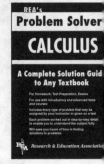

ACCOUNTING
ADVANCED CALCULUS
ALGEBRA & TRIGONOMETRY
AUTOMATIC CONTROL
   SYSTEMS/ROBOTICS
BIOLOGY
BUSINESS, ACCOUNTING, & FINANCE
CALCULUS
CHEMISTRY
COMPLEX VARIABLES
DIFFERENTIAL EQUATIONS
ECONOMICS
ELECTRICAL MACHINES
ELECTRIC CIRCUITS
ELECTROMAGNETICS
ELECTRONIC COMMUNICATIONS
ELECTRONICS
FINITE & DISCRETE MATH
FLUID MECHANICS/DYNAMICS
GENETICS
GEOMETRY
HEAT TRANSFER

LINEAR ALGEBRA
MACHINE DESIGN
MATHEMATICS for ENGINEERS
MECHANICS
NUMERICAL ANALYSIS
OPERATIONS RESEARCH
OPTICS
ORGANIC CHEMISTRY
PHYSICAL CHEMISTRY
PHYSICS
PRE-CALCULUS
PROBABILITY
PSYCHOLOGY
STATISTICS
STRENGTH OF MATERIALS &
   MECHANICS OF SOLIDS
TECHNICAL DESIGN GRAPHICS
THERMODYNAMICS
TOPOLOGY
TRANSPORT PHENOMENA
VECTOR ANALYSIS

*If you would like more information about any of these books,
complete the coupon below and return it to us or visit your local bookstore.*

RESEARCH & EDUCATION ASSOCIATION
61 Ethel Road W. • Piscataway, New Jersey 08854
Phone: (732) 819-8880   **website: www.rea.com**

Please send me more information about your Problem Solver books

Name _____

Address _____

City _____ State _____ Zip _____